SpringerBriefs in GIS

SpringerBriefs in GIS present compact, concise summaries of cutting-edge research, practical applications and visualizations in the use of geographical information systems. At 50 to 125 pages (approx. 20,000 – 50,000) words, SpringerBriefs in GIS provides researchers and practitioners with an innovative venue to present work that might be longer and more complex in scope than a journal article.

This series aims to cover a wide range of topics related to geographical information science and geographical information systems. Potential topics could include: an in-depth case study on the use of GIS to accomplish a specific goal; a guide to an emerging GIS tool, technique or map; a "hot-take" on or snapshot of a current issue that needs to be published as quickly as possible (just 8-12 weeks after a manuscript's delivery and acceptance). Multidisciplinary studies are particularly welcome.

SpringerBriefs are distributed through the same channels as Springer's book content, and are available as physical books and full and chapter-wise eBooks. Both solicited and unsolicited manuscripts are considered for publication in this series. Please send questions and proposals to Zachary Romano, Associate Editor, Earth Science, Environment, and Geography, at Zachary.Romano@Springer.com.

Kousik Das Malakar • Supriya Roy

Mapping Geospatial Citizenship

The Power of Participatory GIS

 Springer

Kousik Das Malakar 🆔
Department of Geography
School of Basic Sciences
Central University of Haryana
Mahendragarh, Haryana, India

Supriya Roy 🆔
Department of Geography
Institute of Humanities & Social Sciences
Visva-Bharati University
Santiniketan, West Bengal, India

ISSN 2367-010X   ISSN 2367-0118   (electronic)
SpringerBriefs in GIS
ISBN 978-3-031-63106-1   ISBN 978-3-031-63107-8   (eBook)
https://doi.org/10.1007/978-3-031-63107-8

This Springer imprint is published by the registered company Springer Nature Switzerland AG
The registered company address is: Gewerbestrasse 11, 6330 Cham, Switzerland

If disposing of this product, please recycle the paper.

Preface

In the era of artificial intelligence, rapid technical breakthroughs, and increased recognition of the importance of geographical information, the convergence of geospatial technology and citizenship has emerged as a critical topic of research. But the challenge is, how can we connect local communities and indigenous people to the geospatial world? Based on this context, this book developed the idea of "Mapping Geospatial Citizenship," employing the power of the Participatory Geographic Information System (PGIS). First and foremost, we must understand what PGIS is and how it relates to social science research. And the answer is that PGIS is a powerful tool for incorporating local people's voices into the GIS systems. This is important in social science research because it provides powerful approaches for exploring, investigating, collecting, and comprehending various socio-spatial phenomena in our socio-ecological systems. And it facilitates research into complex spatial relationships, community dynamics, resource dynamics, and the consequences of policies and actions on local communities.

The book *Mapping Geospatial Citizenship: The Power of Participatory GIS* addresses the transformative potential of PGIS in empowering communities and amplifying their voices by employing geospatial technologies. It consolidates prominent contributions from a range of social science disciplines, including but not limited to anthropology, archaeology, area studies, communication studies, development studies, environmental studies, geography, health studies, media studies, political science, religious studies, rural studies, social policy, social work, socio-ecological studies, sociology, and urban studies. The book emphasizes disciplinary principles, (re)evaluates policy management practices, explores the potential scope for future research, and presents insightful arguments regarding the application of PGIS and community integration. Essentially, the book is dedicated to illustrating four key points:

- Incorporating community perspectives into Geographic Information Systems (GIS).
- Real-world case studies and field narratives on citizens' voice mapping.

- Explains how geospatial thinking can be used to reflect community participation and management.
- It sets itself apart by integrating varied viewpoints from the social sciences and GIScience into its coverage.

The book has been divided into ten interrelated and critical aspect-based chapters connected by three conjunction parts. Part I, *Fundamentals of Geospatial Citizenship and Participatory GIS*, is divided into six chapters. Chapter 1 provides an overview of the introductory section of geospatial citizenship, which will assist us in understanding the concepts, approaches, and dimensions of geospatial citizenship and integrate us to create a mental map on the role of GIS in citizen participation and empowerment, as well as participatory decision-making and collaborative decision-making in the face of the climate crisis in an exploration of space, society, environments, development, and sustainability. Chapter 2 delves into PGIS concepts and techniques, including traditional GIS challenges, community participation and data collection methods, participatory mapping tools and technologies, policy planning, and implementing PGIS approaches for social and spatial justice, among others. Chapter 3 digs into the multiple applications of participatory GIS, including socio-ecological techniques to mapping. In this context, we focus special attention on natural resource management and justice, climate disaster risk reduction and resilience building, urban planning and design, health and social justice, and climate migration policy planning.

Chapter 4 addresses the geographies and socio-spatial ecologies of a societal area using participatory GIS. This journey began with a grasp of the concept of societal spaces, which included geographies and socio-spatial ecologies. Furthermore, it explores community identity, empowering local communities using participatory GIS, and the future of participatory spatial planning. Chapter 5 presents an overview of community cartography in participatory GIS. In this chapter, we go over the concept and approaches of community cartography, participatory mapping techniques, and steps for empowering communities through community cartography, challenges, innovations, planning, socio-spatial justice, and future directions of community cartography. And Chap. 6 declares that GIS is for everyone and discusses the challenges of today's GIS, democratizing PGIS, and the application opportunities of open-source GIS, free tools, web-based GIS, mobile GIS, and crowd-sourced mapping, as well as the promotion of PGIS literacy and future directions.

Part II of the book, *Voices Mapping of Coastal Communities: Field Narratives, Case Studies, and Best Practices of Participatory GIS*, connects the three chapters. Chapter 7 discusses the significance and scope of voices mapping in coastal communities around the coastal World. The importance and scope of community voice mapping for sustainability were primarily emphasized here, as well as the collection of scientific information from global coasts, especially coastal Bengal in India. Chapter 8 reveals a unique and outstanding set of field narratives from the coastal Medinipur and Sundarbans as part of the field inquiry for voices mapping of coastal people. In this regard, we look into human-mangrove conflicts, man-environment linkages, socio-ecological transformation, difficulties and conservation, sustainable thinking in the traditional maritime fishing community, and the extent of climate gap management among climate migrants. Chapter 9 focuses on in-depth case studies and best practices for using PGIS to map coastal community voices in the coastal region of Bengal, India. Therefore, the primary focuses were on mapping indigenous technical knowledge for fish catching, mapping social vulnerability to climate, participatory mapping of fishing grounds, natural resource management using participatory GIS, and modeling vegetation ecologies in relation to climate change. The book's final part delves more into disciplinary principles and social science research scopes as a power of PGIS. In this regard, more than 35 social science fields evaluate and summarize integration facts with PGIS in social science research.

This book highlighted some noteworthy thinking regarding contemporary concerns such as crisis, spatial justice, policy planning, sustainability, and the scope of further research, which is thoroughly discussed and concluded, and each chapter provided some preliminary and enlightening arguments in support of specific points in order to continue thinking about community participation, empowerment, and spatial justice through the GIS systems.

In closing, this book advances scientific knowledge based on participatory GIS among scientists, professionals, researchers, planners, students, and laypeople, while also providing a deeper understanding of community engagement as a geospatial citizen in a social setting. It also encourages various social stakeholders to participate in decision-making and helps planners and authorities build appropriate plans for a region's long-term management and development. As an outcome, it is recommended for everyone interested in geospatial technologies' potential to promote community engagement, social justice, and sustainable development. It provides a roadmap and additional research materials for leveraging the potential of PGIS to transform our perspectives on and engagement with communities, fostering more inclusive and equitable societies. In this context, we, the authors, would like to highlight the following lines:

The creativity that transforms obstacles into possibilities is found in the intricate tapestry of society, where socio-spatial justice, crisis response, and community participation are woven together. Here, participatory GIS acts as a collective map, guiding us through a terrain marked by empowerment and inclusivity.

Mahendragarh, India   Kousik Das Malakar
Santiniketan, India   Supriya Roy

Acknowledgment

We sincerely acknowledge and appreciate the coastal population of West Bengal, India, for their pioneering participation and cooperation during the field investigation.

Kousik Das Malakar and Supriya Roy

Contents

About the Authors

Kousik Das Malakar is a socio-ecological geographer and GIS analyst. He is currently working as a senior research fellow (doctoral) in the Department of Geography, School of Basic Sciences, Central University of Haryana, India. He holds a master's degree in social science geography from Jawaharlal Nehru University (JNU) in New Delhi, and a bachelor's degree in social science geography from Vidyasagar University in Medinipur. In his research area, Malakar has attended over 150 conferences/workshops/seminars/webinars and has published books and research articles in national and international journals. He has received numerous certifications from the Indian Institute of Remote Sensing (IIRS) in Dehradun, the Geological Survey of India (GSI) in Hyderabad, and the National Institute of Disaster Management (NIDM) in New Delhi for his ability to participate in and acquirement of research ideas and knowledge. He is an active journal reviewer and a member of the Human Development and Capability Association (US), the Unequal World (US), and the International Society for Urban Health (US). Climate crisis, socio-ecological systems (governance and security), socio-ecological transformation, coastal society, environments, sustainability, sustainable thinking, disaster studies, and geospatial technologies are among his research interests.

Supriya Roy is a GIS analyst and socio-environmental geographer. She holds a master's degree in social science geography from the Department of Geography, Institute of Humanities & Social Sciences, Visva-Bharati University, Santiniketan, and a double bachelor's degree in social science geography from the University of Gour Banga and Vidyasagar College of Education, West Bengal (India). In her field of research, she has participated in more than 100 workshops/seminars/conferences/webinars and has contributed research articles to both national and international journals. She has been awarded multiple certifications from NIDM (New Delhi) and IIRS (Dehradun) for her active participation in research and the acquisition of knowledge and research ideas. Her research interests encompass a diverse range of topics, including coastal indigenous communities, anthropo-environmental conflicts, climate justice, climate migration, society and politics, participatory policy planning, and the applications of geographic information science.

Abbreviation

C.D. Block	Community Development Block
CC	Climate Change
CC-GIS	Community Cartography with GIS
CC-Tech	Community Cartography Technologies
Cit-GIS	Citizen GIS
C-Map	Community Mapping
EO	Earth Observation
FGDs	Focus Group Discussions
FGs	Fishing Grounds
GC	Geospatial Citizenship
GIS	Geographic Information System
GIScience	Geographic Information Science
GPS	Global Position System
ICT	Information and Communication Technology
ITK	Indigenous Technical Knowledge
NDVI	Normalized Difference Vegetation Index
NGOs	Non-governmental Organizations
OGC	Open Geospatial Consortium
P-GeoTech	Participatory Geospatial Technology
PGIS	Participatory Geographic Information System
PGM	Participatory Geospatial Mapping
PPGIS	Public Participatory Geographic Information System
RS	Remote Sensing
SES	Socio-ecological Systems
SET	Socio-ecological Transformation
TMFC	Traditional Marine Fishing Community
TMFS	Traditional Marine Fishing Society
UNESCO	United Nations Educational, Scientific and Cultural Organization
VGI	Volunteered Geographic Information

Part I
Fundamentals of Geospatial Citizenship and Participatory GIS

Chapter 1
Introduction to Geospatial Citizenship

Geospatial citizenship is about empowering individuals and communities to become stewards of their own landscapes.

Michael F. Goodchild

Abstract The chapter "Introduction to Geospatial Citizenship" provides an in-depth discussion of the concept, various viewpoints, and significance of geospatial citizenship in our increasingly interconnected world. It explores the connections between geography, technology, and civic engagement, highlighting the importance of spatial literacy and appropriate geospatial data use. The chapter first defines geospatial citizenship as people's active participation and responsible engagement in understanding, using, and contributing to the geospatial environment before delving into its characteristics. It emphasizes how geospatial technology improves our understanding of the environment, simplifies decision-making, and promotes long-term growth. In this sense, the author(s) emphasizes the rapid advancement of geospatial technologies such as global positioning systems (GPS), geographic information systems (GIS), and remote sensing. These breakthroughs have radically altered how we collect, process, and comprehend geospatial data, allowing us to make educated decisions about a wide range of issues, including urban planning, environmental management, and disaster response. Moreover, the chapter delves into collaborative decision-making in the geospatial realm, stressing the importance of developing critical thinking and spatial reasoning skills at both individual and community levels. It explores the role of geospatial citizenship in enhancing civic engagement, underscoring how geospatial technologies empower people to influence policies, advocate for their rights, and impact their surroundings. Emphasizing the transformative potential of geospatial citizenship for an inclusive and sustainable society, the chapter concludes by highlighting the need for geospatial literacy, responsible data usage, and active participation for the benefit of local communities and the planet.

Keywords Geospatial environment · Geospatial citizenship · Participatory decision-making · Community participation and empowerment · Geographic information systems

In today's fast-expanding digital world, understanding geospatial citizenship is critical. This chapter examines the core ideas of geospatial citizenship, highlighting its importance in today's society. Geospatial citizenship is fundamentally about engaging with geospatial technology and data in a responsible and informed manner. This chapter lays the groundwork for a thorough examination of the geospatial landscape by instilling civic duty and encouraging the ethical use of location-based information. So, we go on a journey to become responsible geospatial citizens, covering topics such as the importance of spatial awareness and ethical issues in geospatial applications.

Key Points of the Chapter
- Concept and relation between GIS and population.
- Various dimensions of geospatial citizenship and relations with PGIS.
- Thinking about participatory decision-making and empowerment.

1.1   Geospatial Citizenship: Concept and Approaches

1.1.1   Concept

Geospatial citizenship refers to the reasonable and ethical use of geospatial technology and data in an increasingly interconnected society. It includes understanding how location-based information is gathered, used, and shared, as well as the ramifications for individuals, groups, and society as a whole. Geospatial citizenship promotes educated decision-making, environmental stewardship, and social responsibility using geospatial tools such as geographic information systems (GIS) and global positioning systems. It emphasizes the significance of privacy, data security, and equal access to geospatial resources for everyone's benefit. Figure 1.1 depicts the visual representation of the key elements and relationships in geospatial citizenship. As well, we may recall the following:

- Geospatial citizenship involves responsibly using geospatial technology and data and recognizing the rights and responsibilities of individuals and communities in spatial information usage.
- It highlights the significance of spatial literacy, ethical geospatial data use, and active participation in decision-making with geographic information.
- Geospatial citizens understand the socioeconomic, environmental, and governance implications of location-based data and advocate for fair access and ethical behaviours.
- The idea encourages people to actively shape their spatial environment, whether by reporting issues in their area or participating in mitigation activities.

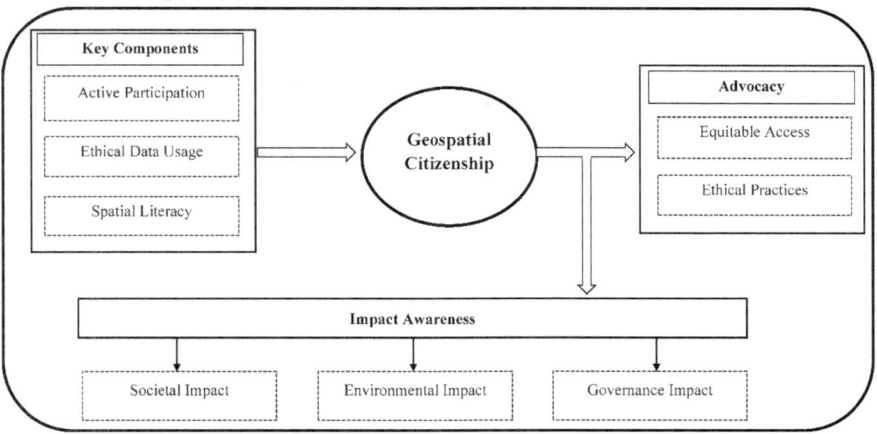

Fig. 1.1 Geospatial citizenship: essential components and connections. (Made by author(s))

1.1.2 Approaches

- *Community participation*: Community participation entails including residents in local geospatial projects and decision-making processes, which fosters a sense of ownership and responsibility for their geographical surrounds.
- *Educational engagement*: Encouraging people to gain geospatial literacy through formal and informal education. This includes teaching the fundamentals of geographic information, spatial reasoning, and responsible data use.
- *Environmental stewardship*: Environmental stewardship entails using geospatial technologies to monitor and resolve environmental issues, empowering citizens to make sustainable decisions and engage in conservation initiatives.
- *Advocacy for access*: Ensuring fair access to geospatial tools and information, bridging the digital divide, and encouraging social inclusion.
- *Ethical awareness*: Promoting ethical considerations in geographic data collection, analysis, and distribution, including privacy, security, and prejudice.
- *Data privacy advocacy*: Educating people about the importance of personal location data privacy and campaigning for open data regulations and practices.
- *Technological innovation*: Encouraging the development and adoption of geospatial technologies that benefit society and adhere to ethical norms.
- *Global citizenship*: Global citizenship entails encouraging a larger perspective on geospatial concerns, acknowledging the interconnection of global challenges, and instilling a sense of duty beyond national borders.

These approaches help to produce responsible and active geospatial citizens who use location-based information to benefit themselves and their communities.

1.1.3 Participatory GIS and Geospatial Citizenship

Participatory GIS (PGIS) and geospatial citizenship are two interconnected concepts that have the potential to transform how societies interact with spatial information, make decisions, and address a variety of challenges, including those related to the environment, urban planning, and social justice. So we need to first understand what PGIS genuinely is.

PGIS is primarily an approach that actively incorporates communities and stakeholders in the spatial data collection, analysis, and decision-making processes. It enables individuals and communities to use geographic information technologies (GIS, GPS, and remote sensing) to address issues specific to their local context. PGIS incorporates local knowledge and skills, making it an effective tool for comprehending complicated spatial challenges and developing context-specific solutions. It frequently encourages collaboration among community members, researchers, government institutions, and nongovernmental organizations (NGOs), thereby encouraging inclusive and democratic decision-making. PGIS can help with a variety of challenges, including land use management, disaster management, resource distribution, urban planning, and environmental conservation.

Relationship between PGIS and geospatial citizenship:

• Participatory GIS (PGIS) puts geospatial citizenship ideals into practice. It contains the idea that individuals and communities should have agency and a voice in the gathering, use, and application of geographical data to address local issues.
• Geospatial citizenship establishes an ethical and responsible framework under which PGIS operates. It assures that participatory geospatial activities protect privacy, adhere to data ethics, and encourage inclusivity and openness.
• PGIS promotes geospatial citizenship by providing individuals and communities with the tools and skills necessary to actively participate in geographical decision-making processes.
• Geospatial citizenship, in turn, encourages the adoption and responsible use of PGIS as a tool for democratizing access to geographic information and promoting more fair and sustainable solutions to societal concerns.

In summary, participatory GIS and geospatial citizenship are two interconnected concepts that enable individuals and communities to actively engage with spatial data and contribute to more inclusive, ethical, and informed decision-making processes. Together, they have the ability to effect beneficial change in a variety of fields, including environmental protection, social justice, and urban planning.

1.2 Dimensions of Geospatial Citizenship: A Legal and Ethical Viewpoint

The following (Fig. 1.2) are the several dimensions of geospatial citizenship:

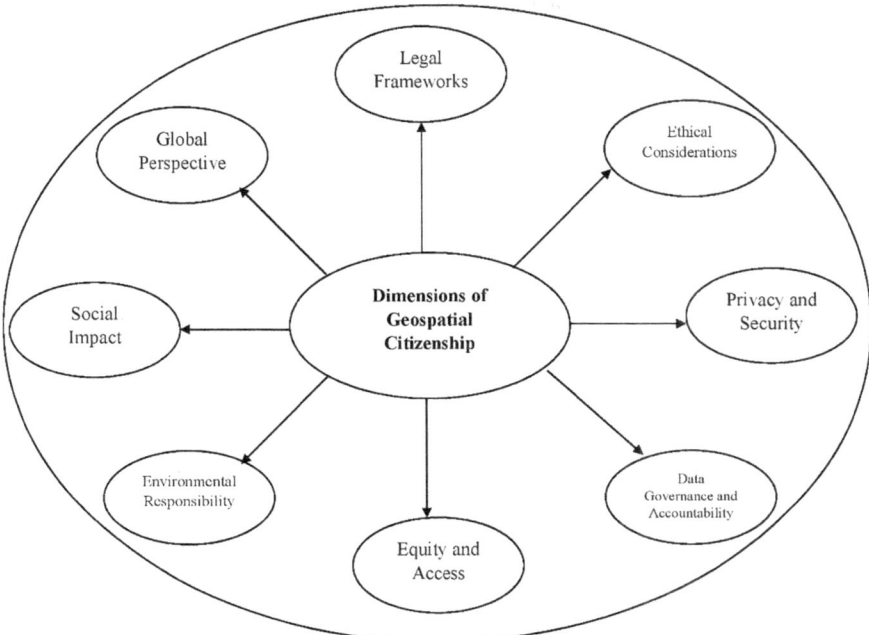

Fig. 1.2 Several dimensions of geospatial citizenship. (Made by author(s))

- *Global perspective*: Geospatial citizenship goes beyond borders, stressing the interdependence of geographical concerns. It emphasizes the ethical obligation to address global issues such as climate change, disaster management, and humanitarian crises.
- *Legal frameworks*: Geospatial citizenship requires adhering to a complicated web of legal restrictions. This dimension emphasizes the need for individuals and companies to understand and adhere to geospatial data laws such as privacy, data protection requirements, and intellectual property rights.
- *Ethical considerations*: Geospatial citizenship stresses ethical behavior when gathering, analyzing, and utilizing location-based data. This dimension emphasizes concepts such as data transparency, informed consent, data accuracy, and reducing harm to individuals and communities.
- *Privacy and security*: Protecting people's location data is critical. Geospatial citizenship necessitates a proactive strategy for protecting sensitive data from unwanted access, breaches, and misuse.
- *Data governance and accountability*: Promoting appropriate data governance and accountability procedures inside businesses and institutions is critical. This component highlights the importance of clear data policies, openness, and means for addressing ethical violations.
- *Equity and access*: This component focuses on the ethical obligation of providing equitable access to geospatial resources and technology. It emphasizes the

need to bridge the digital gap and make geospatial data and tools available to everyone, regardless of socioeconomic background.

- *Environmental responsibility*: Geospatial citizenship includes ethical environmental practices that emphasize the appropriate use of location data for environmental conservation and sustainable development.
- *Social impact*: Understanding the societal impact of geospatial technology is critical. This dimension encourages thoughtful consideration of how location-based information might affect communities, individuals, and vulnerable populations, while also advocating for responsible use.

Therefore, incorporating these features into geospatial citizenship assures that individuals and organizations negotiate the complicated landscape of location-based data with legal compliance, ethical integrity, and a dedication to societal and environmental well-being.

1.3 The Role of GIS in Citizen Participation and Empowerment

Geographic information systems (GIS) play an important role in increasing citizen participation and empowerment in a variety of ways:

- *Spatial data visualization*: GIS empowers individuals to view complex data on maps, making it more accessible and intelligible. This enables individuals to understand local concerns and patterns, resulting in more informed decision-making.
- *Data access and transparency*: GIS platforms enable easy access to government and public data, encouraging transparency and accountability. Citizens have access to information on public services, infrastructure, and land use, allowing them to hold authorities accountable for their activities.
- *Public participation in urban planning*: GIS improves public participation in urban planning processes. Citizens can comment on proposed developments, suggest changes, and participate in collaborative decision-making to ensure that urban planning meets community requirements.
- *Emergency response and disaster management*: GIS plays an important role in catastrophe preparation and response. Citizens can receive real-time information on hazards, evacuation routes, and safe areas, allowing them to make informed decisions during an emergency.
- *Environmental conservation*: GIS provides citizens with the ability to monitor and report environmental changes such as pollution, deforestation, and habitat destruction. This involvement encourages environmental responsibility and contributes to conservation efforts.
- *Healthcare and social services*: GIS can help identify underserved areas and gaps in healthcare and social services. Citizens can fight for increased access to

these programs, which will ultimately contribute to enhanced community well-being.

* *Political engagement*: GIS helps in redistricting and border delineation, ensuring equitable representation. Citizens can get involved in political processes by learning about voting districts and attending elections with more understanding.
* *Community mapping*: With GIS, residents may actively participate in mapping their own communities. This approach involves children in identifying problems, developing solutions, and pushing for change, which increases their sense of ownership and empowerment.

Therefore, GIS technology is a powerful instrument for involving citizens in decision-making processes, promoting transparency, and empowering individuals and communities to actively contribute to the improvement of their communities and society as a whole.

1.4   Geospatial Technologies in Participatory Decision-Making

Geospatial technologies, such as geographic information systems (GIS), global positioning systems (GPS), remote sensing, and interactive mapping tools, have transformed how societies make decisions, especially in terms of participatory governance and community involvement. These technologies enable individuals and communities to participate in decision-making processes in a variety of ways:

* *Data collection and visualization*: Geospatial technologies allow for the collection, organization, and visualization of geographically referenced data. Citizens and government agencies can map different aspects of their surroundings, including land use, infrastructure, environmental quality, and public services. This data visualization simplifies complex information, allowing stakeholders to see obstacles and possibilities.
* *Community mapping*: Citizens can actively help to create maps that reflect their local knowledge and concerns. Residents frequently contribute to community mapping initiatives by identifying risks, unmet needs, and infrastructure gaps. This participatory method promotes a sense of ownership and responsibility in communities.
* *Environmental monitoring*: Geospatial technologies enable citizens and organizations to track changes in the environment. This is particularly useful for monitoring deforestation, pollution, water quality, and habitat conservation. Communities that contribute to environmental data gathering can advocate for conservation and sustainable practices.
* *Regional and urban planning*: GIS can help urban and regional planners visualize development proposals and land use plans. Citizens can submit feedback on

planned projects, ensuring that urban development fits with community interests and contributes to more livable, sustainable communities.

- *Planning for emergency response*: Geospatial technology provides real-time information on the location and magnitude of disasters. This data helps first responders and civilians make informed decisions about evacuation routes and safety precautions, eventually saving lives.
- *Healthcare and social services*: Geospatial tools can help to find the healthcare inequities and service gaps. By mapping healthcare facilities, disease outbreaks, and vulnerable people, communities may fight for better access to healthcare and social services.
- *Political engagement*: Geospatial technologies help to ensure fair political representation through redistricting and boundary demarcation. Citizens can participate in the political process with a better awareness of voting districts and demographics.
- *Infrastructure planning*: GIS helps to plan and manage vital infrastructure such as roads, utilities, and transportation networks. Citizens can report infrastructure faults using mobile apps or online platforms, allowing for more efficient maintenance and repair.
- *Natural resource management*: Geospatial data is used in agriculture and natural resource management to optimize resource allocation, resulting in increased agricultural productivity and sustainable land use practices. Farmers and landowners can make data-driven decisions.
- *Transparency and accountability*: Geospatial technologies encourage transparency and accountability in government. Citizens may track progress and outcomes thanks to publicly accessible maps and data, which hold decision-makers responsible for their actions.

Ultimately, geospatial technologies have been shown to be effective tools for participatory decision-making in a variety of sectors. They allow citizens and communities to actively change their environments, making decisions that are more inclusive, informed, and sensitive to the needs and ambitions of those affected. This transformative power has the ability to build stronger, more egalitarian, and sustainable communities and societies.

1.5 Geospatial Citizenship in the Climate Crisis: Collaborative Decision-Making

The climate crisis is one of the most important global issues of our day, requiring immediate and concerted action from individuals, communities, governments, and organizations. Geospatial citizenship, which emphasizes responsible and informed use of geospatial technologies, is critical for promoting collaborative decision-making to address the climate issue. Here's an in-depth look at how geospatial citizenship might help with collaborative decision-making in this context:

- *Data-driven understanding*: Geospatial technologies allow for the collection, processing, and display of climate-related data such as temperature variations, sea-level rise, and severe weather events. Geospatial residents can actively participate in data-driven climate research, helping to better understand local and global climate trends.
- *Community resilience*: Geospatial citizenship enables residents to map vulnerabilities and analyze climate threats. Communities can use collaborative geospatial technologies to identify regions at risk of flooding, wildfires, or high heat, allowing them to design localized resilience strategies.
- *Disaster preparedness and response*: Geospatial technologies, such as remote sensing and GIS, help with disaster preparedness and response in the face of climate-related disasters. Geospatial residents can help reduce catastrophe risk by reporting real-time data, allowing authorities to make more informed decisions during emergencies.
- *Sustainable urban design*: Sustainable urban design is critical for reducing and adapting to climate change. Geospatial tools help to visualize urban data, and collaborative decision-making platforms enable individuals to contribute ideas on sustainable transportation, green infrastructure, and climate-resilient city design.
- *Conservation and reforestation*: Geospatial technologies help monitor deforestation, land degradation, and biodiversity loss. Geospatial individuals can actively contribute to conservation efforts by reporting illegal logging or finding potential reforestation and habitat restoration sites.
- *Renewable energy planning*: Geospatial data is critical for the planning and implementation of renewable energy projects. Geospatial residents can advocate for and contribute to the development of solar, wind, and hydroelectric energy solutions on a local and regional scale.
- *Environmental advocacy*: Geospatial citizenship can strengthen environmental advocacy activities. Citizens and organizations may raise awareness, generate support, and influence policy for sustainable behaviours by mapping the effects of climate change and environmental deterioration.
- *International collaboration*: Geospatial citizenship has no borders. Geospatial technologies enhance global collaboration, allowing for the sharing of climate data, best practices, and mitigation plans across countries and regions.
- *Equity and inclusion*: Prioritizing fairness and inclusivity is crucial when making collaborative geographical decisions. Geospatial citizens may advance climate justice by ensuring equitable representation of poor communities in climate projects and by avoiding disproportionate impacts on vulnerable people while implementing climate solutions.
- *Monitoring and accountability*: Geospatial technologies allow us to track climate commitments and hold governments and corporations accountable for their climate actions. Geospatial residents can actively monitor carbon reductions, deforestation rates, and progress toward climate goals.

To summarize, collective decision-making within the framework of geospatial citizenship is an essential instrument for addressing climate issues. It enables individuals and communities to actively participate in climate change, offers data-driven insights, and promotes collaboration at the local, national, and global levels. As the climate catastrophe worsens, geospatial citizenship will become increasingly important in encouraging educated, collaborative, and long-term solutions to reduce its effects and provide a more resilient and sustainable future for all.

1.6   Geospatial Citizenship: Social Space, Environment, and Development

Geospatial citizenship is a diversified and revolutionary power that has important implications for social space, environmental sustainability, and development, ultimately achieving spatial literacy (Fig. 1.3). In terms of *social space*, it empowers individuals and communities by increasing spatial awareness and instilling a sense of place-based identity. Citizens can use geospatial tools and technology to actively map and portray their local landscapes, improving their connection to their surroundings. This sense of place promotes community cohesion and civic participation as people become more aware of the distinctive difficulties and opportunities in local social settings. On the other hand, geospatial citizenship has a significant impact on *environmental sustainability*. It enables informed monitoring of natural resources, biodiversity, and environmental changes via data-driven analysis and visualization. This, in turn, allows individuals and groups to actively participate in conservation initiatives, advocate for sustainable land use practices, and help

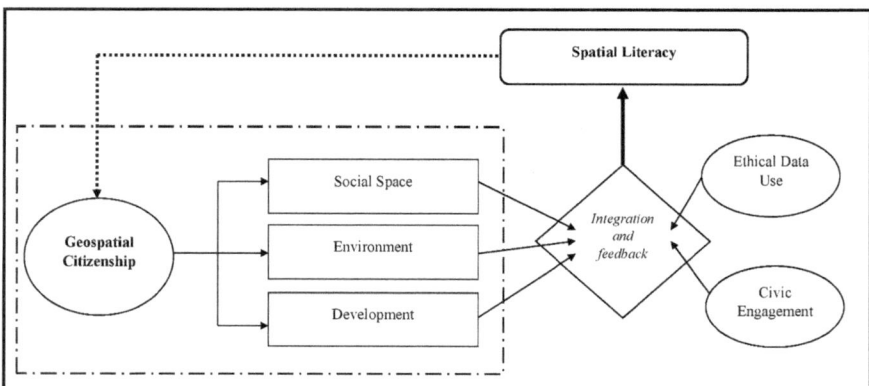

Fig. 1.3 Transformative influence on social space, environmental sustainability, and development, leading to spatial literacy. (Made by author(s))

mitigate environmental concerns such as climate change. Geospatial technologies enable the assessment of the environmental impact of development projects as well as the identification of habitat preservation and restoration zones, fostering harmonious cohabitation between society and nature. Moreover, geospatial citizenship is critical in the context of *development* for making informed decisions. It helps with urban and rural planning by visualizing infrastructure needs, identifying disaster-prone areas, and optimizing resource allocation. Geospatial data also helps with transportation planning, water resource management, and energy distribution, resulting in more efficient and sustainable development practices. Furthermore, geospatial citizenship promotes openness and accountability in development projects, allowing residents to actively participate in community development activities and ensuring that they are in line with local needs and goals.

In conclusion, geospatial citizenship goes beyond simple technical interaction; it reshapes social space by enhancing community relationships and encouraging active participation. Simultaneously, it promotes environmental sustainability by facilitating data-driven conservation and encouraging responsible land use. Finally, it contributes to development initiatives by giving vital geographical insights, resulting in more efficient and fair development processes. Geospatial citizenship serves as a unifying force, bridging the divide between society, the environment, and development and paving the path for a more informed, inclusive, and sustainable future.

1.7 Geospatial Citizenship: Society and Sustainability

Geospatial citizenship and sustainability in society are inextricably intertwined, paving the way for a more responsible and sustainable future. Geospatial citizenship, which focuses on informed and ethical engagement with geospatial technologies and data, supports sustainable practices in a variety of ways. It enables individuals to make data-driven decisions, which contribute to environmental monitoring, resource management, and urban planning in line with sustainability objectives. Citizens can actively participate in conservation initiatives, fight for equitable access to resources, and influence policies that prioritize sustainability by using geospatial tools to enhance community involvement. Furthermore, geospatial technology contributes to catastrophe preparedness and response, increasing societal resilience, which is an important component of sustainability. Geospatial citizenship, in essence, serves as a catalyst for informed, egalitarian, and ecologically responsible decision-making, which is critical to achieving sustainability in today's complex and interconnected world (Fig. 1.4).

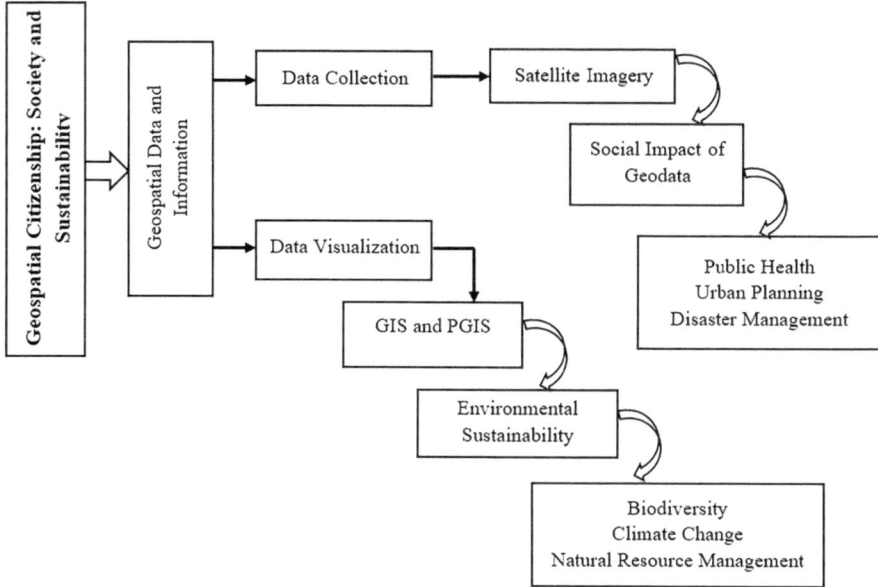

Fig. 1.4 Mapping geospatial citizenship through society and sustainability. (Made by author(s))

1.8 Scope of Research

The future study scope for geospatial citizenship and participatory decision-making in the context of geospatial technologies presents interesting opportunities for investigation and advancement. Some important areas (Fig. 1.5) for research are as follows:

- *User-centric design*: Researchers can focus on creating user-friendly geospatial tools and platforms that appeal to a wide range of people, providing accessibility and inclusion in decision-making processes.
- *Ethical frameworks*: Creating comprehensive ethical frameworks to drive geospatial citizenship and participative decision-making, including data privacy, openness, and accountability.
- *Geo-education*: Investigating the efficacy of geospatial education programs in schools and communities to promote geospatial literacy and responsible citizenship.
- *AI and machine learning integration*: Investigating how artificial intelligence and machine learning may improve geospatial data analysis and help make better decisions, while also taking into account ethical implications and biases.
- *Community empowerment*: Researching techniques to empower underprivileged groups using geospatial technologies, ensuring that all voices are heard in decision-making.

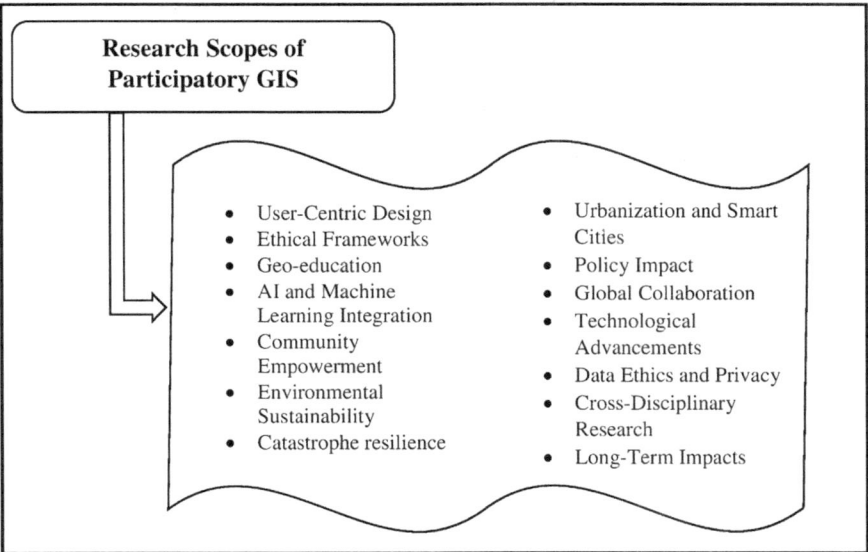

Fig. 1.5 Research scope of participatory GIS. (Made by author(s))

- *Environmental sustainability*: Further research can be conducted into the application of geospatial technology for monitoring and mitigating environmental issues such as climate change, deforestation, and water resource management.
- *Catastrophe resilience*: Investigating how geospatial technology might be used to improve catastrophe preparedness, response, and recovery, with an emphasis on community involvement.
- *Urbanization and smart cities*: This study looks into the use of geospatial technology in sustainable urban planning and smart city development, with a focus on citizen involvement.
- *Policy impact*: Evaluating the impact of geospatial public engagement on local, regional, and national policymaking, as well as analyzing the effectiveness of geospatial-influenced policies.
- *Global collaboration*: Investigating prospects for worldwide collaboration in geospatial citizenship and participatory decision-making, particularly in response to global concerns such as pandemics, migration, and climate adaptation.
- *Technological advancements*: Researchers can look into upcoming geospatial technologies like augmented reality, blockchain, and quantum computing, as well as their potential impact on participatory decision-making and geospatial citizenship.
- *Data ethics and privacy*: Continuously examining the ethical implications and privacy concerns related to geographic data gathering, sharing, and use, as well as implementing appropriate data governance mechanisms.
- *Cross-disciplinary research*: It promotes interdisciplinary research that integrates geospatial technology with fields such as social sciences, public policy, and environmental studies to achieve a holistic comprehension of their impact on decision-making processes.

- *Long-term impacts*: Examining the long-term effects of geospatial citizenship programs on sustainability, resilience, and community well-being, as well as establishing best practices for ongoing engagement.

By exploring these research directions, scholars and practitioners can propel the development of geospatial citizenship and participatory decision-making, leading to the advancement of societies that are more informed, equitable, and sustainable.

1.9   Conclusion and Arguments

In the end, it may be acknowledged that geospatial citizenship is a transformative force that has a significant impact on development, environmental sustainability, and social space. Through improving spatial awareness, fortifying relationships in local settings, and encouraging civic engagement, it empowers people individually and in communities. Geospatial tools facilitate data-driven environmental monitoring and help make well-informed decisions to solve today's urgent issues, such as biodiversity loss and climate change. Moreover, geospatial citizenship encourages accountability and openness in development, which results in more effective and fair procedures. However, in order to fully realize geospatial citizenship's promise for achieving a more informed, inclusive, and sustainable future, challenges such as data privacy, access disparities, and ethical considerations must be carefully addressed.

In light of the climate catastrophe and socio-ecological issues, geospatial citizenship and participatory GIS also provide a critical and creative approach to decision-making. By including citizens in the gathering and analysis of geospatial data, we not only increase access to information but also promote a sense of community responsibility. However, one major difficulty is ensuring the ethical use of geospatial technology, particularly in terms of data protection and security. Innovative technologies, such as decentralized and blockchain-based data management, can address these issues by allowing individuals to actively participate to sustainability initiatives while still protecting their rights and privacy in an increasingly interconnected world.

Further Reading

Abbot, J., Chambers, R., Dunn, C., Harris, T., de Merode, E., Porter, G., Townsend, J., & Weiner, D. (1998). Participatory GIS: Opportunity or oxymoron? *PLA Notes, 33*, 27–34. http://www.iapad.org/wp-content/uploads/2015/07/participatory_gis_opportunity_or_oxymoron.pdf

Brown, G. (2012). An empirical evaluation of the spatial accuracy of public participation GIS (PPGIS) data. *Applied Geography, 34*, 289–294. https://doi.org/10.1016/j.apgeog.2011.12.004

Chrisman, N. R. (1999). What does 'GIS' mean? *Transactions in GIS, 3*, 175–186. https://doi.org/10.1111/1467-9671.00014

Dunn, C. E. (2007). Participatory GIS—A people's GIS? *Progress in Human Geography, 31*(5), 616–637. https://doi.org/10.1177/0309132507081493

Gold, C. M. (2006). What is GIS and what is not? *Transactions in GIS, 10*, 505–519. https://doi.org/10.1111/j.1467-9671.2006.01009.x

Harmsworth, G. (1998). Indigenous values and GIS: A method and a framework. *Indigenous Knowledge and Development Monitor, 6*(3), 1–7.

Kahila-Tani, M., Broberg, A., Kyttä, M., & Tyger, T. (2016). Let the citizens map- public participation GIS as a planning support system in the Helsinki master plan process. *Planning Practice & Research, 31*(2), 195–214. https://doi.org/10.1080/02697459.2015.1104203

Kocaman, S., Saran, S., Durmaz, M., & Kumar, A. (Eds.). (2022). *Citizen science and geospatial capacity building*. MDPI-Multidisciplinary Digital Publishing Institute. ISBN 978-3-0365-3714-6. https://doi.org/10.3390/books978-3-0365-3714-6

Lafreniere, D., Weidner, L., Trepal, D., Scarlett, S. F., Arnold, J., Pastel, R., & Williams, R. (2019). Public participatory historical GIS. *Historical Methods: A Journal of Quantitative and Interdisciplinary History, 52*(3), 132–149. https://doi.org/10.1080/01615440.2019.1567418

Petch, J., & Reeve, D. E. (1999). *GIS, organisations and people: A socio-technical approach*. CRC Press. ISBN: 9780748406531.

Shin, E. E., & Bednarz, S. W. (Eds.). (2018). *Spatial citizenship education: Citizenship through geography*. Routledge. ISBN: 978-1138056442.

Williams, C., & Dunn, C. E. (2003). GIS in participatory research: Assessing the impact of landmines on communities in North-West Cambodia. *Transactions in GIS, 7*, 393–410.

Chapter 2
Understanding of Participatory GIS: Concepts and Techniques

Maps are the tools of empowerment, allowing communities to visualize their challenges and collaborate on solutions.

Jack Dangermond

Abstract This chapter provides an in-depth look at participatory geographic information systems (PGIS) and their essential concepts and techniques. It addresses the integration of participatory techniques with GIS technology, highlighting the significance of inclusive decision-making processes and community participation in spatial planning and resource management. The chapter opens by defining participatory GIS as a means of involving local populations in data collection, analysis, and decision-making processes that integrate geographic information systems (GIS) with participatory approaches. It emphasizes the shift away from traditional, top-down approaches and toward more inclusive and collaborative practices that empower communities while also promoting social justice and management. Furthermore, the chapter covers community participation in data gathering and discusses data gathering methods by offering the skills and information to actively participate in decision-making processes about their community's local environment. In addition, the chapter discusses the many methods and technologies for participatory mapping and innovation. The chapter finishes by emphasizing participatory GIS' transformative potential in encouraging community engagement, social inclusion, and sustainable development. It emphasizes the value of collaboration, trust-building, and ongoing learning while applying participatory methodologies. Inclusive GIS is regarded as a potent tool for democratizing spatial data and enabling more egalitarian and inclusive decision-making processes. Overall, this chapter offers an in-depth discussion and community arguments about participatory GIS and its fundamental principles and methodologies. It emphasizes the significance of inclusive and participatory approaches in GIS, which allow communities to actively participate in decision-making processes, contribute local knowledge, and promote social and spatial justice.

Keywords Participatory GIS · Community participation · Social and spatial justice · Local populations · Sustainable development

© The Author(s), under exclusive license to Springer Nature Switzerland AG 2024   19
K. D. Malakar, S. Roy, *Mapping Geospatial Citizenship*, SpringerBriefs in GIS,
https://doi.org/10.1007/978-3-031-63107-8_2

In this chapter, we delve into participatory geographic information systems (GIS). We'll look at the fundamental concepts and strategies that support this collaborative approach to geographic data management. Participatory GIS enables communities to actively participate in mapping, decision-making, and issue resolution, improving our understanding and interaction with the environment. Through a succinct analysis of its concepts, tools, and applications, we hope to provide you with firm knowledge of how this unique method is transforming the landscape of geographic information, bridging the gap between technology and community empowerment.

Key Points of the Chapter
- Basic concepts of the PGIS.
- Studying the tools and technologies for participatory mapping.
- It covers community participation approaches as well as promoting social justice and management.

2.1 Participatory GIS: Concept and Principles

2.1.1 Concept

Participatory GIS (PGIS) is a collaborative approach to geospatial data collection and decision-making (Fig. 2.1). Mapping and evaluating spatial data requires the participation of communities, stakeholders, and professionals. PGIS empowers local knowledge, allowing communities to share their perspectives and concerns in decision-making processes. It promotes openness, encourages community ownership, and aids in the resolution of complicated issues such as land use, natural resource management, and urban planning. PGIS tools span from GPS devices to web-based platforms, making them a versatile tool for connecting technology, local knowledge, and sustainable development.

2.1.2 Principles

The following principles guide the ethical and effective use of PGIS for informed decision-making and community empowerment:

- *Inclusivity*: Ensure that different stakeholders, especially marginalized groups, have a say in data collection and decision-making.
- *Capacity building*: Provide training and tools to help participants improve their GIS skills and understanding.
- *Transparency*: Maintain openness in data exchange and decision-making processes to build confidence among participants.

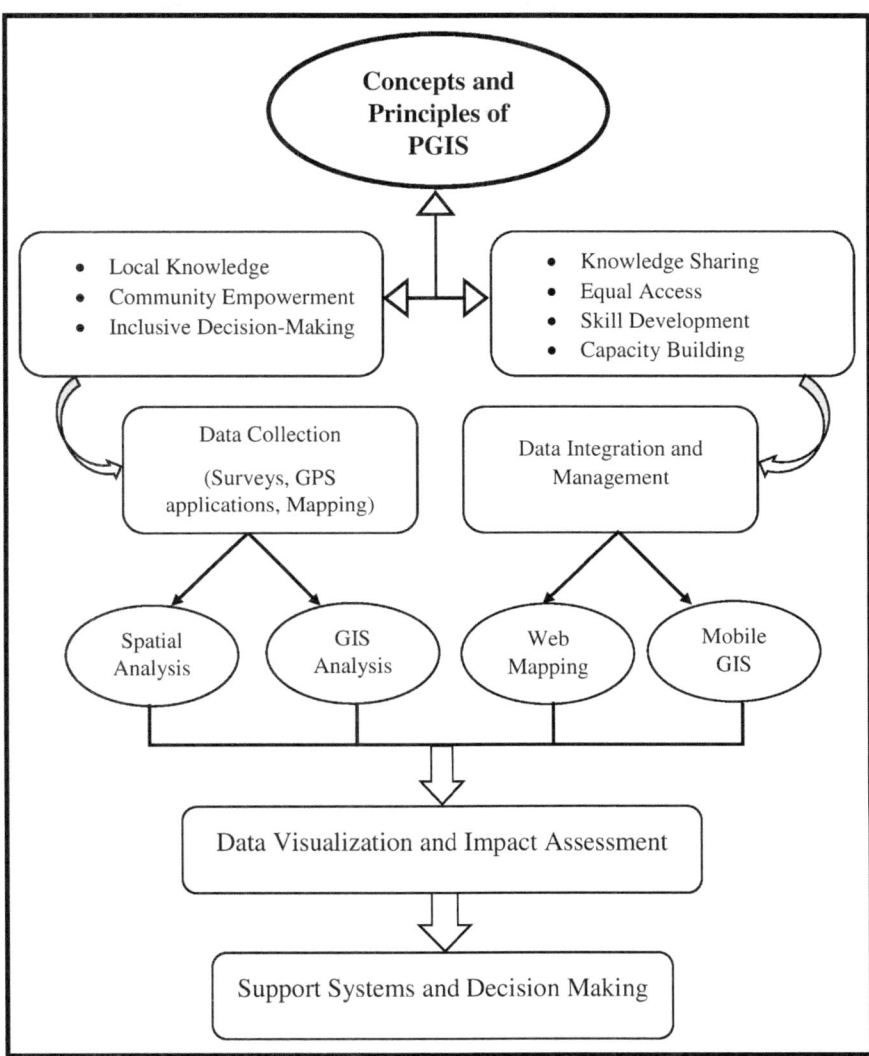

Fig. 2.1 Concepts and principles of participatory GIS. (Made by author(s))

- *Accessibility*: Make GIS tools and information available to all participants, regardless of their technical background.
- *Flexibility*: Customize PGIS tools and processes to meet the specific needs and circumstances of any community or project.
- *Ethical considerations*: In collecting and distributing data, consider privacy and data security considerations, as well as cultural sensitivities.
- *Empowerment*: Empower communities by allowing them to actively participate in mapping and analysis, thereby increasing local knowledge.

- *Sustainable engagement*: Encourage long-term community involvement to ensure ongoing data updates and problem solutions.

2.1.3   Applications, Advantages, and Disadvantages of Participatory GIS

Applications of Participatory GIS
- *Natural resource management*: It helps to monitor and manage forests, water resources, and animals by including local communities in data gathering and conservation initiatives.
- *Disaster management*: PGIS aids disaster preparedness and response by facilitating rapid data sharing and real-time information interchange among stakeholders.
- *Urban planning*: Communities can use PGIS to handle urban concerns such as transportation, infrastructure development, and environmental quality, resulting in more inclusive and sustainable communities.
- *Land use planning*: PGIS enables communities and authorities to plan for sustainable land use in a collaborative manner, taking into account local and environmental concerns.
- *Healthcare*: PGIS can monitor disease outbreaks and healthcare requirements, allowing for targeted interventions in neglected areas.

Advantages of Participatory GIS
- *Community empowerment*: It enables local communities to actively participate in decision-making processes, instilling a sense of ownership and accountability.
- *Transparency*: Promotes transparency and accountability by making GIS data accessible to all stakeholders.
- *Local knowledge*: Uses valuable local knowledge to enhance the accuracy and usefulness of spatial data.
- *Cost-effectiveness*: Involving communities in data collection can lower the cost of field surveys while increasing data accuracy.
- *Conflict resolution*: Provides a forum for debate and negotiation to aid in the resolution of land and resource use disputes.

Disadvantages of Participatory GIS
- *Technical challenges*: Community members' lack of technical competence might make effective involvement difficult.
- *Data quality*: When nonexperts collect data, it can be difficult to ensure accuracy and reliability.
- *Bias and inequality*: There may be biases in data gathering and power imbalances among stakeholders, resulting in unequal representation.

- *Resource-intensive*: Implementing PGIS may involve a significant amount of time, effort, and money, including training and capacity building.
- *Privacy concerns*: Sharing personal or sensitive information in PGIS may cause privacy and security issues.

2.1.4 Integration of Participative Tactics with GIS Technology

The combination of participatory approaches and geographic information systems (GIS) technology is a powerful synergy with far-reaching ramifications. This strategy entails involving communities in data collection and analysis, which enriches GIS databases with local knowledge and insights. Residents actively contribute to the identification of critical areas, resources, and issues through participatory mapping exercises, which fosters a sense of ownership and empowerment. GIS technology is important because it provides tools for organizing, visualizing, and analyzing cooperatively collected data.

This integration increases openness, strengthens decision-making, and allows for more effective resource management, making it an important strategy in areas such as land use planning, disaster management, and natural resource conservation. Nonetheless, successful adoption necessitates overcoming technological hurdles, ensuring data quality, and mitigating any biases in order to reap the benefits of this strong combination of GIS and community participation.

2.1.5 Participatory GIS for Inclusive Decision-Making Processes

Participatory geographic information systems (PGIS) are extremely important in promoting inclusive decision-making processes. By incorporating communities and stakeholders in data gathering and analysis, PGIS ensures that a wide range of perspectives are heard. This approach strengthens underrepresented groups and increases their representation in land use, resource management, urban planning, and other decision-making processes. PGIS fosters openness by making spatial data available to everyone, bridging the divide between technical specialists and local knowledge. Its applications include conflict resolution, improved service delivery, and tackling social and environmental justice issues. In essence, PGIS acts as a catalyst for more fair and informed decision-making, making it an essential tool for fostering inclusive and sustainable communities.

2.1.6 Community Participation in Spatial Planning and Resource Management

Participatory geographic information systems (PGIS) are critical tools for increasing community engagement in spatial planning and resource management. By incorporating people and stakeholders in data gathering, analysis, and decision-making, PGIS allows communities to actively influence their surroundings. This inclusive approach instils a sense of ownership and responsibility in participants, ensuring that local knowledge and concerns are factored into planning. PGIS uses include sustainable land use planning, natural resource management, disaster preparedness, and urban development. It enables more informed and equitable decision-making, increases community resilience, and promotes resource sustainability. In essence, PGIS connects technology and community interaction, resulting in more effective and community-centered spatial planning and resource management (Fig. 2.2).

2.2 Traditional GIS Challenges and PGIS

Traditional GIS comes with drawbacks such as data inaccessibility, low community engagement, and the potential exclusion of local knowledge. Participatory GIS (PGIS) solves these issues by integrating communities in data collection, increasing transparency, and bridging the knowledge gap between professionals and citizens. PGIS improves data accuracy by using local insights, empowering communities, and promoting inclusive decision-making. It converts GIS from a top-down tool to a collaborative platform, making it a useful technique in a variety of sectors such as urban planning, resource management, and disaster response. In essence, PGIS is a more people-focused and responsive version of standard GIS.

Traditional geographic information systems have long been a cornerstone of spatial data processing, but they present inherent problems that participatory GIS aims to address.

Traditional GIS Challenges
- *Data inaccessibility*: GIS data is usually kept by experts or government agencies, limiting public access.
- *Limited community engagement*: Traditional GIS typically excludes community input from data collection and decision-making.
- *Data accuracy*: It may lack local expertise and on-the-ground insights, resulting in inaccuracies.
- *Expert-centric*: It relies primarily on technical expertise, which may alienate nonexperts.
- *Top-down approach*: Decision-making is frequently top-down, with little regard for community opinions.

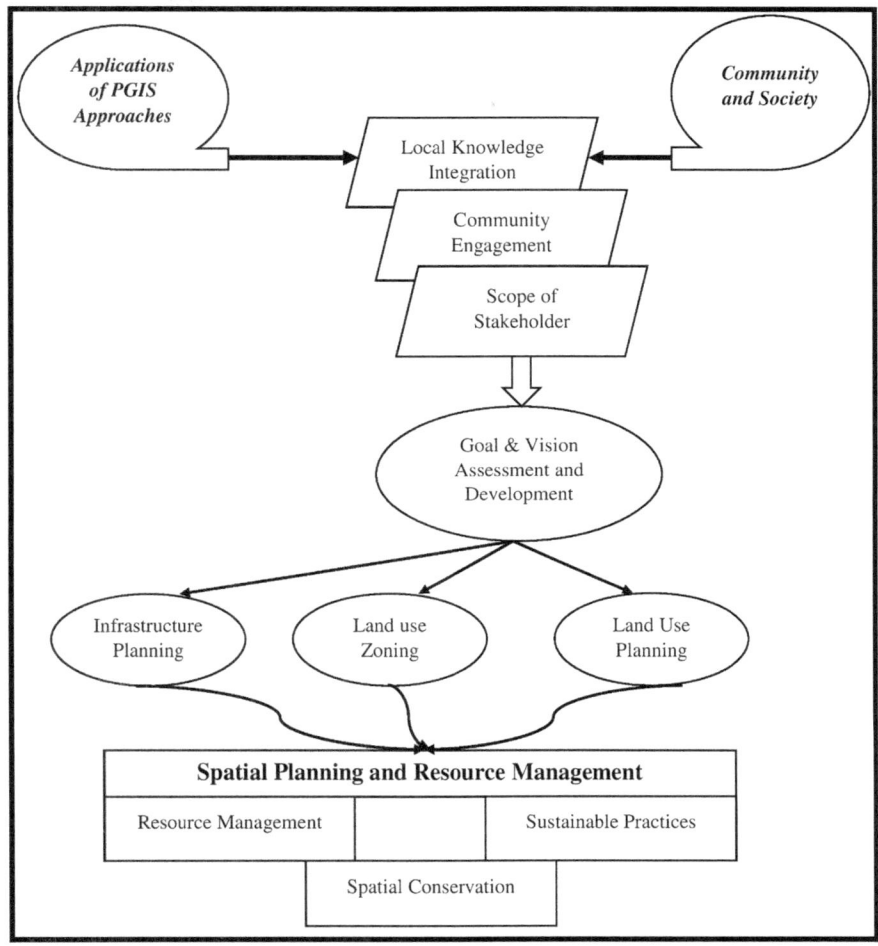

Fig. 2.2 Spatial planning and resource management through PGIS. (Made by author(s))

Solutions Are in PGIS

- *Community involvement*: PGIS actively involves communities in data collection, giving them the opportunity to participate in the process.
- *Transparency*: It encourages open access to GIS data and decision-making, hence building confidence.
- *Local knowledge*: PGIS incorporates local experience, which improves data quality and relevance.
- *Empowerment*: Communities take an active role in changing their environs.
- *Inclusive decision-making*: PGIS makes certain that a wider range of perspectives are addressed in planning and resource management.

In conclusion, PGIS provides a more inclusive, transparent, and community-driven approach to geographical data analysis and decision-making, overcoming many of traditional GIS's drawbacks.

2.3   Community Participation and Data Collection Methods

Participatory GIS uses a variety of data collection approaches to actively engage people in gathering and managing spatial data (Fig. 2.3). This collaborative strategy encourages community engagement and empowers local stakeholders. The following are some significant data collection methods utilized in PGIS and their significance:

- *Community surveys*: Surveys are a core PGIS tool. Communities are surveyed to learn about their needs, preferences, and spatial knowledge. This information can help with decisions about urban development, resource management, and disaster preparedness.

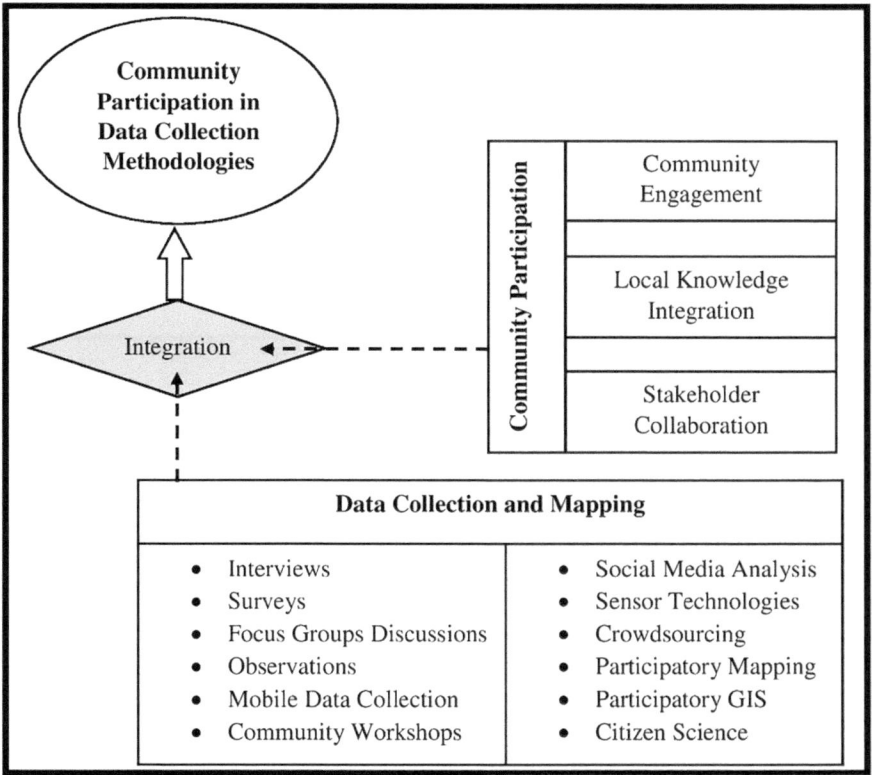

Fig. 2.3 Community participation in data collection methodologies. (Made by Author(s))

- *Focus group discussions (FGD)*: During these discussions, community members share their perspectives and concerns about their surroundings. The data acquired can be utilized to identify key areas for intervention or development.
- *Community mapping*: Community members actively participate in mapping exercises, which involve marking key areas, resources, and concerns on physical maps or digital platforms. These maps incorporate local knowledge, making them useful for land-use planning and natural resource management.
- *GPS and mobile apps*: PGIS frequently uses GPS devices or mobile applications, allowing community members to readily capture geospatial data. This strategy is very beneficial for disaster response and environmental monitoring.
- *Photovoice*: Participants use cameras or smartphones to take pictures of locations and topics that are important to them. These images serve as a visual depiction of community issues, which can influence decision-makers in areas such as infrastructure development.
- *Participatory 3D modeling (P3DM)*: It involves communities creating physical or digital three-dimensional models of their environments, providing for a tactile depiction of their surroundings. This is particularly beneficial for catastrophe risk reduction and resource management.

Significance
- *Local knowledge integration*: PGIS approaches ensure that local knowledge, which is sometimes disregarded in standard GIS, is included in spatial data, increasing its accuracy and usefulness.
- *Empowerment*: These strategies empower communities by allowing them to participate in data collection and decision-making processes.
- *Transparency*: Data acquired using PGIS technologies is visible and accessible to all stakeholders, which fosters trust and accountability.
- *Inclusivity*: These strategies promote the engagement of marginalized groups, resulting in a more inclusive approach to decision-making.
- *Improved decision-making*: PGIS data gathering methods help to make better informed and community-driven decisions in a variety of fields, including urban planning and disaster management.

To summarize, community participation using PGIS data gathering methods transforms geographic data management by prioritizing local knowledge, inclusion, and transparency, ultimately leading to more responsive and sustainable decision-making processes.

2.4 Tools and Technologies for Participatory Mapping

Participatory geographic information systems (PGIS) use a variety of tools and technologies to involve communities in mapping and data collection (Fig. 2.4). These tools empower local stakeholders, improve data accuracy, and allow for

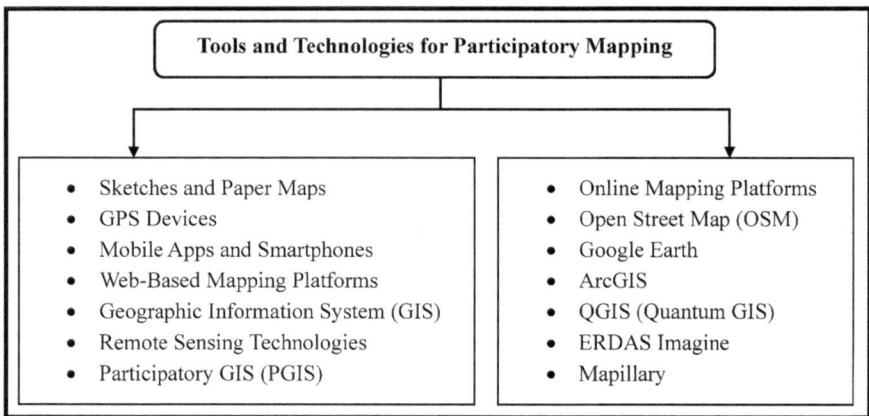

Fig. 2.4 Tools and technologies for participatory mapping. (Made by author(s))

collaborative decision-making. Here are some essential tools frequently used in PGIS:

- *Paper maps and sketches*: Simple paper maps and sketches enable community members to annotate, draw, and identify key areas and features. These low-tech instruments are available to most communities and are frequently utilized in workshops and gatherings.
- *Community mapping workshops*: Hands-on workshops, led by experts or groups, introduce communities to mapping tools and techniques. These seminars provide community members with training and support so that they can effectively participate in mapping efforts.
- *Geographic information systems (GIS) software*: GIS software such as QGIS, ArcGIS Online, and open-source solutions offer powerful tools for producing, editing, and analyzing spatial data. These tools can be customized for community use with the proper training.
- *Global positioning system (GPS) devices*: Handheld GPS devices or smartphone apps with GPS capabilities allow community members to collect exact position data for mapping purposes. GPS tools are extremely useful for resource mapping and environmental monitoring.
- *Smartphone apps*: Mobile applications such as OpenStreetMap and Mapbox empower individuals to contribute to open-source maps, update information, and add geospatial data using their smartphones. These apps foster real-time data collection and community engagement.
- *Participatory 3D modeling (P3DM) tools*: Specialized P3DM kits include modeling materials and software to assist communities in developing physical or digital three-dimensional models of local environments. P3DM is particularly useful for catastrophe risk reduction and urban planning.
- *Web-based mapping platforms*: Platforms such as Google Earth, Mapbox, and Esri's ArcGIS Online provide user-friendly interfaces that enable communities

to collaboratively generate and share maps via the internet, encouraging greater engagement.
- *Drones and aerial imagery*: Aerial imaging and drones may collect high-resolution photographs and data from above, allowing communities to track land use changes, assess environmental conditions, and plan infrastructure projects.

To sum up, the diverse range of participatory mapping tools in PGIS enables communities to actively participate in data collection, promotes data accuracy, and fosters inclusive, transparent, and informed decision-making processes in a variety of domains, from environmental conservation to urban planning.

2.5 Application of PGIS in Decision-Making Processes

Participatory geographic information systems (PGIS) have a significant impact on decision-making processes across multiple domains:

- *Urban planning*: PGIS allows communities to actively engage in decisions about urban development, such as infrastructure projects and land use planning. This inclusive strategy promotes the development of more sustainable and community-centered cities.
- *Natural resource management*: PGIS enables local stakeholders to help manage natural resources such as forests, water bodies, and agricultural land. It contributes to sustainable resource use and conservation.
- *Disaster risk reduction*: In disaster-prone locations, PGIS helps with hazard mapping, evacuation plans, and early warning systems. Communities may better prepare for and respond to catastrophes by leveraging local knowledge and data.
- *Healthcare planning*: PGIS can monitor disease outbreaks, healthcare institution locations, and healthcare infrastructure requirements. This data enables targeted healthcare interventions and resource allocation.
- *Land tenure and property rights*: PGIS can aid with land tenure concerns by mapping property borders and recording land ownership. It has the ability to resolve land disputes and grant legal recognition to marginalized populations.
- *Environmental conservation*: PGIS is useful for mapping biodiversity, ecological hotspots, and habitat conservation activities. It aids in making informed decisions to protect the environment and endangered animals.
- *Infrastructure development*: PGIS influences decisions about road construction, water supply, and sanitation, ensuring that infrastructure projects meet local demands and conditions.
- *Social and cultural heritage preservation*: PGIS aids in the documentation of cultural heritage places and traditional knowledge, thereby helping to preserve and conserve them.

- *Community empowerment*: Most importantly, PGIS improves community participation by allowing residents to actively participate in decision-making processes, establishing a sense of ownership and responsibility.

In conclusion, PGIS improves decision-making by incorporating local knowledge, boosting openness, and encouraging inclusivity. It closes the gap between specialists and communities, resulting in better-informed, equitable, and sustainable decisions.

2.6 PGIS Approaches in Climate Crisis Policy Planning

Participatory geographic information systems (PGIS) play an important role in tackling climate change through informed policy planning (Fig. 2.5), such as the following:

- *Disaster preparedness*: PGIS contributes to climate adaptation by mapping susceptible infrastructure and building early warning systems. This allows communities to better prepare for extreme weather occurrences and changing climate circumstances.
- *Community-led climate mapping*: PGIS helps communities map climate-related vulnerabilities, such as flood-prone areas or sea level rise zones. This localized knowledge helps to shape climate adaptation strategies and disaster risk reduction programs.
- *Ecosystem conservation*: PGIS supports mapping and monitoring key ecosystems such as forests and wetlands, allowing policymakers to make more informed decisions on conservation and restoration activities that sequester carbon and increase resilience.
- *Renewable energy siting*: PGIS helps to locate ideal places for renewable energy projects, taking into account local perspectives and environmental implications. This promotes the shift to renewable energy sources.
- *Carbon footprint analysis*: Communities utilize PGIS to measure and visualize their carbon footprints, which leads to grassroots campaigns to reduce emissions and promote sustainable behaviours.
- *Climate education and awareness*: PGIS technologies aid climate education by visually depicting climate data and impacts, raising awareness, and encouraging behavior change at the community level.

Therefore, incorporating PGIS into climate crisis policy planning improves data accuracy, encourages community participation, and indicates that climate policies are effective and socially equitable. These ideas help to create more resilient and sustainable answers to the challenges posed by climate change.

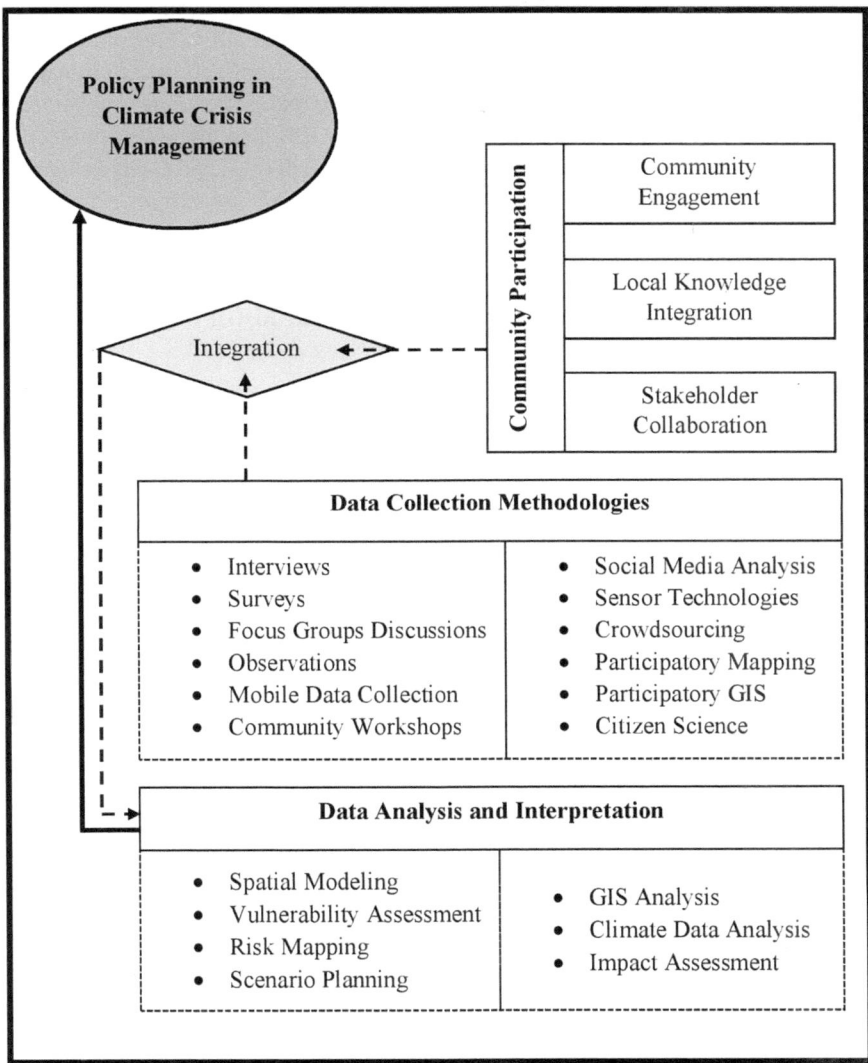

Fig. 2.5   PGIS approaches in climate crisis policy planning. (Made by author(s))

2.7   PGIS Approaches: Impact on Social Space, Environment, and Development

Participatory geographic information systems (PGIS) have a revolutionary impact at the nexus of social space, the environment, and development. These approaches enable communities to actively shape their surroundings by incorporating local knowledge into geographical data. By mapping social places, PGIS promotes inclusive urban planning and infrastructure development, ensuring that the built environment meets community requirements. Simultaneously, PGIS promotes environmental protection by allowing communities to map and manage natural resources, thus contributing to sustainable development. The synergy between social space, environment, and development in PGIS encourages equitable and informed decision-making, strengthens community resilience, and fosters a harmonious balance between human activity and the environment.

2.8   PGIS Approaches, Society, and Sustainability

Participatory geographic information systems (PGIS) techniques establish an important link between society and sustainability. By actively engaging communities in mapping and decision-making processes, PGIS enables individuals and groups to take responsibility for their environmental and socioeconomic problems. This empowerment promotes a sense of responsibility, resulting in more sustainable activities and policies. PGIS also helps society understand its interactions with the environment, providing insights into resource management, land use, and catastrophe preparedness. It enhances the social fabric by encouraging inclusive, transparent, and participatory approaches to addressing sustainability issues. Ultimately, PGIS bridges the gap between society and sustainability, paving the way for more resilient and egalitarian societies that value the planet's long-term health.

2.9   Implementing PGIS Approaches for Social and Spatial Justice in Society

Participatory geographic information systems (PGIS) techniques are useful in promoting social and spatial justice in society. PGIS contributes to the correction of historical injustices in land tenure, resource allocation, and urban planning by actively incorporating marginalized communities and stakeholders in mapping and decision-making. These initiatives give underrepresented groups a say in designing their environments, resulting in more equitable land use policies, infrastructure development, and access to resources. PGIS also highlights geographical inequities and environmental injustices, enabling evidence-based advocacy and policy changes to address issues such as pollution, housing inequalities, and unequal access to services. In summary, PGIS acts as a catalyst for social and spatial justice implementation, encouraging fairness, inclusivity, and a higher quality of life for all members of society.

2.10   Scope of Research

The field of participatory GIS is constantly evolving, providing several prospects to future research for the research community. Some examples are as follows:

- *Community empowerment*: Looking into the empowerment outcomes of PGIS activities and how they affect community resilience and well-being.
- *Methodological advancements*: Research can focus on developing innovative PGIS tools and procedures to make data collection more accessible, accurate, and user-friendly for a wide range of communities.
- *Indigenous knowledge systems*: Explore how PGIS might be used to document and conserve indigenous knowledge and cultural heritage.
- *Integration with emerging technologies*: Looking at how PGIS may use emerging technologies such as artificial intelligence, blockchain, and augmented reality to improve community involvement and data analysis.
- *Global challenges*: Researchers can investigate how PGIS methodologies can help address urgent global issues like biodiversity conservation, disaster risk reduction, and public health concerns.
- *Impact assessment*: Impact assessment is the process of evaluating the long-term social, economic, and environmental implications of PGIS programs in order to better understand their effectiveness and sustainability.
- *Ethical considerations*: Addressing ethical concerns about data privacy, security, and fair representation of underrepresented groups in PGIS processes.
- *Scalability*: Researchers can investigate techniques for expanding PGIS programs to target larger geographic areas and global concerns like climate change and urbanization.
- *Policy integration*: Investigating how PGIS might be included in government policies and decision-making processes to promote inclusive and transparent governance.
- *Cross-disciplinary collaboration*: Encouraging collaboration among PGIS practitioners, social scientists, environmental researchers, and policymakers to address complex societal and environmental challenges.

Therefore, future research in PGIS has the potential to advance the field's techniques, ethical considerations, and effects on society and sustainability, ultimately leading to more inclusive and successful community-driven approaches to geographic data management and decision-making.

2.11   Conclusion and Arguments

In the end, knowledge of participatory GIS highlights its critical role in transforming the landscape of geographic data management and decision-making. By actively engaging communities and stakeholders in data collection and analysis, PGIS empowers individuals, promotes inclusion, and improves data accuracy. Its applications are numerous, ranging from urban planning and natural resource management to catastrophe preparedness and cultural preservation. The future of PGIS research holds great opportunities for methodological breakthroughs, ethical considerations, and cross-disciplinary collaboration, highlighting its importance in tackling critical socioeconomic and environmental challenges. In summary, PGIS is a transformative paradigm that connects technology and community engagement, paving the path for more sustainable and equitable societies.

In closing, PGIS principles and methodologies are critical for solving the climate catastrophe, socioeconomic difficulties, and sustainability by encouraging community engagement, integrating multiple knowledge systems, and ensuring more equitable and effective decision-making processes. This strategy is critical for ensuring a sustainable future in the face of urgent global concerns.

Further Reading

Allen, B. L. (2018). Strongly participatory science and knowledge justice in an environmentally contested region. *Science, Technology & Human Values, 43*(6), 947–971. https://doi.org/10.1177/0162243918758380

Babelon, I., Ståhle, A., & Balfors, B. (2017). Toward cyborg PPGIS: Exploring socio-technical requirements for the use of web-based PPGIS in two municipal planning cases, Stockholm region, Sweden. *Journal of Environmental Planning and Management, 60*(8), 1366–1390. https://doi.org/10.1080/09640568.2016.1221798

Bijker, R. A., & Sijtsma, F. J. (2017). A portfolio of natural places: Using a participatory GIS tool to compare the appreciation and use of green spaces inside and outside urban areas by urban residents. *Landscape and Urban Planning, 158*, 155–165. https://doi.org/10.1016/j.landurbplan.2016.10.004

Brown, G. (2012). An empirical evaluation of the spatial accuracy of public participation GIS (PPGIS) data. *Applied Geography, 34*, 289–294. https://doi.org/10.1016/J.APGEOG.2011.12.004

Corbett, J., & Keller, P. (2006). An analytical framework to examine empowerment associated with participatory geographic information systems (PGIS). *Cartographica: The International Journal for Geographic Information and Geovisualization, 40*(4), 91–102. https://doi.org/10.3138/J590-6354-P38V-4269

Fagerholm, N., Raymond, C. M., Olafsson, A. S., Brown, G., Rinne, T., Hasanzadeh, K., et al. (2021). A methodological framework for analysis of participatory mapping data in research, planning, and management. *International Journal of Geographical Information Science, 35*, 1848. https://doi.org/10.1080/13658816.2020.1869747

Haklay, M., & Tobon, C. (2003). Usability evaluation and PPGIS: Towards a user-centred design approach. *International Journal of Geographical Information Science, 17*, 577–592.

Mitchell, A. (1997). Zeroing. In *Geographic information systems at work in the community.* Environmental Systems Research Institute, U.S. ISBN: 978-1879102507.

Obermeyer, N. J. (1998). The evolution of public participation GIS. *Cartography and Geographic Information Systems, 25*, 65–66.

Ramasubramanian, L. (2010). *Geographic information science and public participation* (Advances in geographic information science). Springer Berlin. https://doi.org/10.1007/978-3-540-75401-5. ISBN: 978-3-540-75400-8.

Rambaldi, G., Kyem, A. P. K., McCall, M., & Weiner, D. (2006). Participatory spatial information management and communication in developing countries. *Electronic Journal on Information Systems in Developing Countries, 25*(1), 1–9. https://doi.org/10.1002/j.1681-4835.2006.tb00162.x

Sanderson, E., & Kindon, S. (2004). Progress in participatory development: Opening up the possibility of knowledge through progressive participation. *Progress in Development Studies, 4*, 114–126.

Sieber, R. (2006). Public participation geographic information systems: A literature review and framework. *Annals of the Association of American Geographers, 96*(3), 491–507. https://doi.org/10.1111/j.1467-8306.2006.00702.x

Tripp, C. (2011). *Evaluation of a public participatory GIS tool: A public planning case study.* VDM Verlag. ISBN: 978-3639359145.

Chapter 3
Applications of Participatory GIS: A Socio-ecological Approaches and Mapping

In the hands of communities, GIS becomes a tool for resilience, fostering a deeper connection between people and their environments.

Jane Goodall

Abstract This chapter explores the various uses of participatory PGIS in socio-ecological systems. It demonstrates how the integration of participatory techniques and GIS technology may help to understand and address complex socio-ecological concerns while also supporting sustainable development and community resilience. The chapter opens with an introduction to socio-ecological systems and mapping that highlights the interdependence of social and ecological components. When tackling environmental and socioeconomic concerns, it underlines the need to consider the linkages between human activities, ecosystems, and natural resources. The author delves into the several uses of participatory GIS in socio-ecological systems. Additionally, look for how PGIS might be used for community-based natural resource management, participatory land-use planning, and sustainable agriculture. The chapter demonstrates how participatory approaches improve awareness of local environments, encourage community engagement, and support informed decision-making in each application. The chapter also looks at the use of participatory GIS in catastrophe risk reduction and climate change adaptation, as well as urban planning and design. It highlights the importance of incorporating local communities in hazard mapping, vulnerability assessments, urban planning, design, and adaptive strategy development. Participatory techniques allow communities to share their knowledge and experiences, increasing resilience and lowering disaster risks. The chapter focuses on the significance of participatory GIS in social justice and activism. It highlights how the integration of GIS technology with participatory methods can support marginalized communities in asserting their rights, documenting socio-environmental injustices, and advocating for policy changes. Participatory mapping and community-led data collection are effective strategies for amplifying marginalized voices and improving social fairness. Furthermore, the chapter argues that the obstacles and issues connected with using participatory GIS in socio-ecological contexts are significant.

Keywords Socio-ecological system and mapping · Participatory land use planning · Participatory GIS in disaster risk reduction · Sustainable development · Participatory GIS in socio-environmental justice

In this chapter, we are going to explore the numerous and dynamic uses of participatory GIS within a socio-ecological framework. We investigate the interaction of society and the environment, namely, how PGIS enables people to actively participate in mapping and decision-making processes that solve complex socio-ecological issues. From mapping endangered habitats to measuring the socioeconomic implications of environmental policy, we demonstrate how PGIS can bridge the gap between technology and community-driven sustainability. This chapter presents a rich tapestry of possibilities in which PGIS might help to build more resilient, equitable, and environmentally sustainable communities.

Key Points in the Chapter
- Interaction between the socio-ecological system and participatory approach.
- Application of PGIS in the socio-ecological domains.
- Policy planning insights.

3.1   Socio-ecological Mapping and Participatory GIS

Socio-ecological mapping, when combined with participatory GIS, provides a dynamic way to understand and tackle complex environmental and social concerns. It recognizes the complex relationship between human society and the natural environment, emphasizing active community participation. This approach differs from typical GIS in that it documents human activities, land use, and sociocultural contexts in addition to mapping ecological aspects. It allows communities and stakeholders to actively participate in data collection and decision-making, bringing valuable local knowledge to the process. Socio-ecological mapping contributes to resource management, policy development, climate adaption, and environmental justice efforts by offering comprehensive insights into the interactions between people and nature. Ultimately, it functions as a transformative instrument for developing more sustainable and resilient socio-ecological systems.

Socio-ecological mapping within the context of participatory GIS is a strong tool for connecting the social and ecological components of our world. The following provides a detailed overview:

- *Understanding the socio-ecological context*: Socio-ecological mapping begins with the recognition that human civilizations and the natural environment are inextricably interwoven. It seeks to represent this connection by depicting how human actions affect and are influenced by ecological systems.
- *Community engagement*: PGIS concepts are central to socio-ecological mapping, with an emphasis on active community participation. Local citizens and

stakeholders participate in data collection, analysis, and interpretation, bringing significant knowledge to the process.

- *Mapping human-environment interactions*: Socio-ecological mapping differs from typical GIS in that it documents human activities, land use patterns, and social structures in addition to mapping environmental elements. It emphasizes the dynamic relationship between humans and nature.
- *Identifying vulnerabilities*: These maps aid in detecting weaknesses in socio-ecological systems. For example, mapping can indicate areas prone to environmental dangers like flooding and erosion, as well as how they affect local residents.
- *Resource management*: Socio-ecological mapping contributes to the sustainable management of natural resources. Communities can map resource locations, evaluate their status, and plan for conservation or responsible usage.
- *Policy development*: Socio-ecological maps can help policymakers make more informed decisions concerning land use, resource allocation, and environmental protection. These maps provide a comprehensive overview of the repercussions of policy decisions.
- *Environmental justice*: Socio-ecological mapping can highlight environmental injustices, such as the disproportionate impact of pollution or climate change on underserved groups. This data can be used to advocate for more egalitarian policies.
- *Climate change adaptation*: It aids in climate change adaptation by identifying climatic vulnerabilities and community resilience. This information aids in the development of focused strategies for responding to changing environmental conditions.
- *Cultural and indigenous knowledge*: Socio-ecological mapping can combine cultural and indigenous knowledge, maintaining traditional practices and providing insights into sustainable resource management.
- *Ecosystem services*: These maps can analyze and depict ecosystem services like water purification and carbon sequestration, underscoring the importance of healthy ecosystems to human well-being.

Lastly, socio-ecological mapping within the context of participatory GIS provides a comprehensive picture of our complex world, stressing the interconnection of social and environmental systems. It empowers communities, informs policy choices, and promotes sustainable practices, making it an essential tool for solving today's critical socioeconomic and environmental concerns.

3.2   Environmental Resource Management and Justice Through Participatory Mapping

Participatory mapping is essential for achieving environmental justice and sustainable resource management. It prioritizes communities and stakeholders in environmental resource decision-making, instilling a sense of empowerment and accountability. Participatory mapping helps to gain a better knowledge of the relationship between human activities and the environment by mapping natural resources, ecosystems, and environmental vulnerabilities. It not only promotes equitable resource allocation and sustainable management, but it also exposes environmental injustices, focusing attention on underprivileged populations who are disproportionately affected by environmental deterioration. The transparency and inclusivity of participatory mapping promote policy changes and advocacy initiatives, paving the path for more equitable environmental practices, healthier ecosystems, and a more just society.

Participatory mapping, used within the context of environmental resource management (ERM), is an effective method for improving environmental justice and sustainable resource usage. Here's an in-depth look:

- *Community empowerment*: Participatory mapping allows local communities and stakeholders to actively participate in the management of environmental resources. By involving locals in data gathering, analysis, and decision-making, they develop a sense of ownership and accountability.
- *Resource identification and assessment*: Communities can map and appraise natural resources like forests, bodies of water, and agricultural land. This knowledge is crucial for long-term resource management, preventing overexploitation and degradation.
- *Environmental justice*: Participatory mapping identifies environmental injustices, such as the disproportionate impact of pollution or resource depletion on underprivileged groups. This information can help to promote advocacy for more equitable resource allocation and environmental control.
- *Ecosystem preservation*: Mapping important ecosystems and biodiversity hotspots contributes to their preservation. This is critical for preserving ecological services, protecting endangered species, and addressing climate change.
- *Disaster preparedness*: Mapping vulnerable areas and disaster hazards helps to improve disaster preparedness and risk reduction. Communities can devise methods to protect people and property in the face of environmental threats.
- *Policy influence*: Participatory maps can help policymakers make more informed decisions about land use, conservation, and resource allocation. This contributes to more sustainable and equitable environmental policies.
- *Conflict resolution*: In cases of resource conflicts or land tenure issues, participatory mapping offers a transparent and evidence-based method for conflict resolution, promoting peaceful coexistence.
- *Cultural and indigenous knowledge*: Participatory mapping can incorporate cultural and indigenous knowledge, thereby maintaining traditional practices and insights into sustainable resource management.
- *Public understanding*: The mapping process improves public knowledge about environmental challenges and the need for sustainable resource management while also encouraging responsible behavior and community engagement.

In conclusion, participatory mapping in environmental resource management is an effective strategy for attaining environmental justice and sustainable resource usage. It empowers communities, informs policy decisions, and promotes equitable and responsible environmental behaviours, resulting in a healthier world and fairer societies.

3.3 Participatory GIS in Climate Disaster Risk Reduction and Resilience Building

When combined with socio-ecological approaches, participatory geographic information systems (PGIS) can play a critical role in climate disaster risk reduction and resilience development. This junction provides a dynamic strategy that not only engages communities in comprehending climate concerns but also highlights the complex relationships between human civilizations and ecological systems. PGIS allows for active engagement in hazard mapping, resource evaluation, and adaptive planning, which are all critical components of resilience-building activities. By combining social and ecological data, PGIS promotes holistic thinking, which leads to more effective and context-specific catastrophe risk mitigation measures. It encourages communities to take responsibility for their resilience, promotes fair responses, and emphasizes the importance of environmental protection in mitigating vulnerability to climate-related disasters. In essence, PGIS within a socio-ecological framework is an effective tool for promoting climate resilience and improving catastrophe risk reduction initiatives. Participatory GIS has emerged as an important tool for climate disaster risk reduction and resilience building, especially when combined with a socio-ecological approach. The following points discuss an overview of the topic:

- *Socio-ecological understanding*: PGIS integrates social and ecological data to enhance understanding of climate threats, recognizing the interconnectedness of human and environmental systems and their influence on disaster outcomes.
- *Community engagement*: PGIS promotes local community involvement in mapping and understanding climate vulnerabilities, hazards, and resources, recognizing it as essential for effective climate disaster risk reduction and resilience building.
- *Hazard mapping*: Communities use PGIS to map areas susceptible to climate-related disasters like floods, hurricanes, and wildfires, facilitating preparedness and response efforts, ultimately saving lives and minimizing damage.
- *Resource mapping*: PGIS assists in mapping crucial resources like water sources, food supplies, and medical facilities, playing a vital role in understanding resource availability for the development of resilient disaster responses.
- *Community-based early warning systems*: PGIS advocates for community-based early warning systems, enabling residents to contribute real-time data and observations that enhance the timeliness and effectiveness of disaster warnings.
- *Environmental conservation*: PGIS helps to map and conserve natural ecosystems that provide critical functions such as flood control and climate regulation. Preserving these ecosystems helps to reduce disaster risk.
- *Social equity*: A socio-ecological approach to PGIS ensures that climate resilience and catastrophe risk reduction activities are fair. Vulnerable and marginalized communities are given the attention and resources they require to overcome climate-related difficulties.

- *Adaptive planning*: PGIS enables communities to create adaptable plans that take into account both social and ecological concerns. This guarantees that resilience-building tactics are both contextually appropriate and long-lasting.
- *Policy influence*: Participatory mapping and data generated through PGIS offer compelling evidence for informed policy decisions regarding climate disaster risk reduction and resilience building.

In brief, including PGIS in a socio-ecological framework is critical for addressing the complex difficulties posed by climate disasters. This integration empowers communities, fosters resilience, champions environmental sustainability, and encourages equitable responses, leading to a more adaptable and resilient society in the face of climate change.

3.4   Urban Planning and Design: Applications of Participatory GIS

Participatory GIS used in urban planning and design within a socio-ecological context provides novel ideas for sustainable, resilient cities. By involving communities in mapping and decision-making processes, PGIS ensures that urban development meets both social and environmental goals. This approach facilitates the identification of green spaces, ecosystem services, and climate vulnerabilities in urban environments, thereby promoting environmental protection and improving community well-being. PGIS empowers residents to influence their urban environment, addressing issues like equal resource access, affordable housing, and disaster preparedness. By acknowledging the interplay of social and ecological processes in urban settings, PGIS contributes to the development of inclusive, environmentally conscious, and resilient cities capable of thriving amidst urbanization and climate change.

Participatory GIS has transformed urban planning and design by combining socio-ecological perspectives. The following are substantial discussions about the transformative intersection:

- *Sustainable urban design*: PGIS helps identify green spaces, sustainable transit routes, and low-impact development opportunities. It guarantees that urban design follows ecological principles, thereby decreasing cities' environmental footprint.
- *Community engagement*: PGIS helps urban citizens actively shape their cities. By integrating people into mapping and decision-making, they gain a sense of control over urban development.
- *Ecosystem services mapping*: PGIS is useful for mapping urban ecosystem services like urban woods, which provide air purification and temperature regulation. This knowledge is critical to improving urban liability and resilience.

- *Disaster preparedness*: PGIS facilitates the development of community-based early warning systems and disaster response plans, thereby increasing urban resilience in the face of climate-related emergencies.
- *Climate resilience*: Using PGIS, communities can identify climate vulnerabilities in urban areas, enabling targeted climate adaptation initiatives and disaster risk reduction.
- *Affordable housing and equitable access*: The PGIS shows discrepancies in housing, infrastructure, and resource access, motivating urban planners to address concerns such as affordable housing, clean water, and sanitation, resulting in more equitable development.
- *Green infrastructure*: PGIS guides the development of green infrastructure networks in cities, which improve biodiversity, reduce urban heat islands, and encourage sustainable storm water management.
- *Urban health and well-being*: PGIS allows for the mapping of healthcare services, pollution sources, and access to green spaces, which helps with urban health planning and well-being.
- *Cultural preservation*: Using PGIS, urban people may map and safeguard cultural heritage sites and historic neighborhoods, ensuring their preservation in the face of growing urbanization.
- *Policy influence*: Participatory maps and data generated by PGIS provide evidence for informed urban policies and development plans that prioritize sustainability, resilience, and social equality.

In conclusion, the integration of PGIS and socio-ecological approaches in urban planning envisions cities as dynamic, interconnected systems. This collaboration enhances community strength, promotes environmental sustainability, reinforces resilience, and cultivates inclusive, liveable urban environments better equipped to address the challenges of urbanization and climate change.

3.5 Geospatial Citizenship for Health and Social Justice

Geospatial citizenship is an important notion in socio-ecological systems for increasing human health and social justice. It acknowledges the connection between the environment, society, and individual well-being. Using geospatial technologies and data, we can map and evaluate complex interactions between our surroundings, social systems, and health outcomes (Fig. 3.1). This understanding allows us to better address environmental inequalities and health disparities. Geospatial citizenship promotes active environmental stewardship and advocacy for fair access to resources and opportunities. In its ultimate form, it aims to create a more equitable and sustainable future by fostering a sense of responsibility toward our world and fellow citizens through this holistic approach. From the perspective of socio-ecological insights, geospatial citizenship is a dynamic framework that combines civic duty and geographic knowledge to advance social justice and human health. This idea

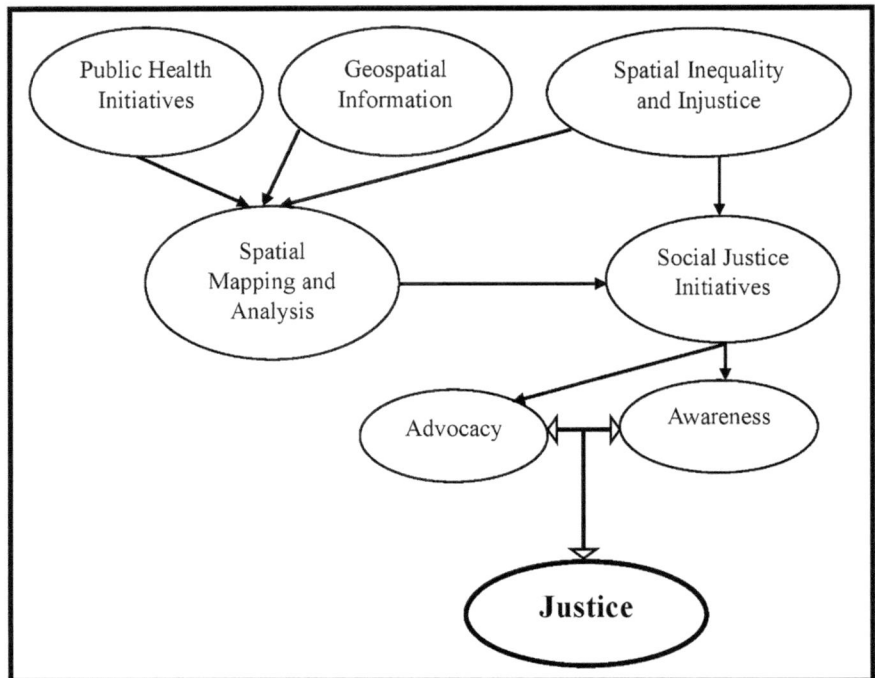

Fig. 3.1 Geospatial citizenship for health and social justice. (Made by author(s))

acknowledges the complex relationships between our environment, society, and personal well-being and the necessity of comprehending these relationships in order to address today's issues. This is a thorough examination of geospatial citizenship within socio-ecological frameworks with respect to human health and social justice, such as the following:

- *Interconnectedness of ecosystems and health*: Socio-ecological perspectives highlight how the environment and human health are interdependent. These complicated relationships, such as the effect of air quality on respiratory health or access to green areas on mental well-being, can be mapped and evaluated using geospatial techniques and data.
- *Environmental inequalities and health disparities*: Disparities in resource availability and environmental conditions can be found through geospatial analysis. By emphasizing how underprivileged people frequently face the brunt of environmental deterioration and its health effects, it helps identify "environmental justice" issues.
- *Citizen engagement and empowerment*: Geospatial citizenship encourages people to actively use geospatial technology and data. It allows people to participate in environmental monitoring, data gathering, and advocacy, establishing a sense of responsibility for their surroundings.

- *Equity and access*: Advocates of geospatial citizenship push for equal access to technology and geographic data. It provides those without opportunity with the resources and information they need to successfully address environmental and health issues.
- *Policy advocacy*: Leveraging geospatial data is crucial for guiding evidence-based policymaking. Geospatial citizens often play a substantial role in advocating for legislation that promotes social justice, public health, and sustainable development.
- *Community resilience*: Geospatial citizenship helps communities establish resilience strategies by knowing their geographical vulnerabilities and strengths.
- *Cross-sectoral collaboration*: Geospatial citizenship connects environmental scientists, public health professionals, urban planners, and policymakers. It promotes interdisciplinary collaboration in addressing difficult socio-ecological issues.
- *Global perspective*: There are no boundaries between local and global citizenship. It encourages a global viewpoint by enabling people and institutions to keep an eye on and react to environmental and public health emergencies, such as disease outbreaks and climate change.
- *Data privacy and ethics*: With geospatial technologies gathering substantial personal data, geospatial citizenship involves addressing data privacy and ethical data use, guaranteeing the respect of individuals' rights.

In summary, geospatial citizenship is a vital component of socio-ecological approaches, providing a comprehensive framework for solving modern concerns in human health and social justice. It makes it possible for people to actively engage with their surroundings, support just laws, and strive toward a more equitable and sustainable society that puts the needs of both people and the environment foremost.

3.6   Climate Migration Policy Planning and PGIS

The incorporation of participatory GIS techniques for climate migration policy planning within a socio-ecological system is critical in dealing with the mounting issues of climate-induced migration. Here's how PGIS can be used in this situation:

- PGIS enables communities to map their own vulnerabilities and identify climate threats. Participatory mapping exercises allow communities to describe regions prone to flooding, drought, or other environmental risks, assisting policymakers in understanding the scope of the problem.
- PGIS can help track climate change-related migration patterns. Community-generated maps can show where individuals are relocating in response to environmental challenges, assisting authorities in anticipating and planning for anticipated influxes of climate migrants.
- Policymakers can use PGIS data to better allocate resources for infrastructure development and services in areas impacted by climate-related migration. This

guarantees that the appropriate support mechanisms are in place in the receiving communities.

- PGIS encourages community participation in decision-making processes. It gives affected people a say in climate migration policy, ensuring that their specific needs and viewpoints are acknowledged.
- Real-time PGIS data can help emergency responders during climate-related disasters by identifying safe evacuation routes and locating temporary shelters for displaced people.
- Climate-induced migration can occasionally result in resource disputes. PGIS can assist in identifying possible conflict hotspots and supporting conflict resolution initiatives by providing transparent and reliable data.
- PGIS data can be used as evidence in legal processes concerning climate-induced migration and human rights breaches. It has the potential to help establish climate migrants' rights and advocate for their protection.
- PGIS can help nearby regions or nations affected by climate migration collaborate more effectively. Sharing spatial data and information can result in coordinated efforts to manage and help displaced populations.
- PGIS enables the incorporation of climate migration issues into long-term resilience plans. It aids in the identification of regions that may become uninhabitable as a result of climate change, as well as the development of managed retreat or adaptation programs.
- PGIS data can be regularly updated to track the efficacy of climate migration policies and assess their influence on populations and ecosystems, allowing for adaptive policy changes.

Lastly, it may be stated that integrating PGIS approaches into climate migration policy planning ensures that these policies are based on local knowledge, address the specific issues encountered by impacted communities, and increase resilience in the face of climate-induced migration. It also encourages a more inclusive and fair approach to addressing this complex socio-ecological challenge.

3.7 Scope of Research

The scope of future studies on the uses of participatory GIS in socio-ecological methods and mapping is large and dynamic, with multiple routes for exploration and advancement. The following are a few intriguing areas for future research:

- Investigate how PGIS might help communities build resilience in the face of environmental changes. This could entail evaluating the efficacy of PGIS-based community-driven adaptation techniques and their effects on socio-ecological systems.
- Investigate strategies for incorporating traditional ecological information held by indigenous and local groups into PGIS platforms. This study can help to bridge the gap between scientific and indigenous perspectives, creating a more comprehensive understanding of ecosystems.

- Evaluate the role of PGIS in climate change mitigation initiatives, such as afforestation and reforestation projects. Determine how PGIS might aid in the mapping of carbon sequestration potential and sustainable land-use planning.
- Investigate the use of PGIS to map and value ecological services. This can help policymakers make more informed judgments about land use planning and conservation activities.
- Examine how PGIS can help reduce health inequalities and environmental injustices. Research might concentrate on mapping environmental health risks and their effects on vulnerable groups.
- Examine the use of PGIS in urban planning, focusing on green infrastructure. Examine how it can be used to map green areas, urban heat islands, and environmentally friendly transit choices.
- Examine how PGIS might help with biodiversity conservation initiatives, such as tracking wildlife numbers, mapping important habitats, and evaluating the success of protected areas.
- Examine the integration of PGIS into the policymaking and decision-making processes across different governmental levels. Analyze its effect on policy alignment with socio-ecological objectives.
- Discuss data privacy and ethical issues with PGIS, especially as they pertain to the gathering and exchange of sensitive data. Provide best practices and recommendations for managing data responsibly.
- Examine how cutting-edge technologies like blockchain, augmented reality, and machine intelligence might be used in PGIS to improve data security, accessibility, and accuracy.
- Study efficient ways to ensure fair access and participation by teaching communities and stakeholders about PGIS methodologies.
- Encourage interdisciplinary research to successfully solve complex socio-ecological concerns by combining knowledge from public health, ecology, geography, sociology, and other domains.

3.8 Conclusion and Arguments

To sum up, the utilization of participatory GIS in socio-ecological methods and mapping presents a diverse range of study options that are constantly expanding. In addition to enabling communities to take an active role in social and environmental decision-making, PGIS promotes a more comprehensive comprehension of intricate problems, including biodiversity preservation, health inequalities, and climate change.

Upcoming research in this area should focus on building community resilience, including indigenous knowledge, and addressing pressing concerns such as urban planning and climate change mitigation. As technology advances and interdisciplinary collaboration expands, PGIS remains an effective tool for developing fair and sustainable socio-ecological systems.

Further Reading

Bernard, E., Barbosa, L., & Carvalho, R. (2011). Participatory GIS in a sustainable use reserve in Brazilian Amazonia: Implications for management and conservation. *Applied Geography, 31*(2), 564–572. https://doi.org/10.1016/j.apgeog.2010.11.014

Brown, G., & Fagerholm, N. (2015). Empirical PPGIS/PGIS mapping of ecosystem services: A review and evaluation. *Ecosystem Services, 13*, 119–133. https://doi.org/10.1016/j.ecoser.2014.10.007

Brown, G., & Raymond, C. M. (2014). Methods for identifying land use conflict potential using participatory mapping. *Landscape and Urban Planning, 122*, 196–208. https://doi.org/10.1016/j.landurbplan.2013.11.007

Brown, G., & Webber, D. (2011). Public participation GIS: A new method for national park planning. *Landscape and Urban Planning, 102*, 1–15. https://doi.org/10.1016/j.landurbplan.2011.03.003

Corbett, J. (2009). *Good practices in participatory mapping: A review prepared for the international fund for agricultural development.* International Fund for Agricultural Development IFAD, Rome. Available at: https://www.ifad.org/documents/38714170/39144386/PM_web.pdf/7c1eda69-8205-4c31-8912-3c25d6f90055

Hasanzadeh, K. (2022). Use of participatory mapping approaches for activity space studies: A brief overview of pros and cons. *GeoJournal, 87*(Suppl 4), 723–738. https://doi.org/10.1007/s10708-021-10489-0

Kahila-Tani, M., Kytta, M., & Geertman, S. (2019). Does mapping improve public participation? Exploring the pros and cons of using public participation GIS in urban planning practices. *Landscape and Urban Planning, 186*, 45–55. https://doi.org/10.1016/j.landurbplan.2019.02.019

McCall, M. K., & Minang, P. A. (2005). Assessing participatory GIS for community-based natural resources management: Claiming community forests in Cameroon. *The Geographical Journal, 171*(4), 340–356. https://doi.org/10.1111/j.1475-4959.2005.00173.x

Mere-Roncal, C., Cardoso Carrero, G., Chavez, A. B., Almeyda Zambrano, A. M., Loiselle, B., Veluk Gutierrez, F., et al. (2021). Participatory mapping for strengthening environmental governance on socio-ecological impacts of infrastructure in the Amazon: Lessons to improve tools and strategies. *Sustainability, 13*(24), 14048. https://doi.org/10.3390/su132414048

Scully-Engelmeyer, K. M., Granek, E. F., Nielsen-Pincus, M., & Brown, G. (2021). Participatory GIS mapping highlights indirect use and existence values of coastal resources and marine conservation areas. *Ecosystem Services, 50*, 101301. https://doi.org/10.1016/j.ecoser.2021.101301

Thompson, D., Lindsay, F. E., Davis, P. E., & Wong, D. W. (1997). Towards a framework for learning with GIS: The case of urban world, a hypermap learning environment based on GIS. *Transactions in GIS, 2*(2), 151–167. https://doi.org/10.1111/j.1467-9671.1997.tb00023.x

Chapter 4
Geographies and Socio-spatial Ecologies of a Societal Space: A Journey into Participatory GIS

In the age of geospatial technology, understanding the socio-spatial ecologies of our society is the key to informed and empowered citizenship.

Michael Batty

Abstract The chapter "Geographies and Socio-spatial Ecologies of a Societal Space: A Journey into Participatory GIS" presents an in-depth appraisal of the geographies and socio-spatial ecologies inside a societal space, with a focus on the function of participatory GIS. It takes the reader on a tour into the realm of participatory GIS, revealing its importance in comprehending the intricate relationships between people, places, and the environment. The chapter begins by laying the foundation for an assessment of societal space, highlighting the interdependence of ecological, social, and geographical issues. It digs into geographies and socio-spatial ecologies, which relate to the dynamic interactions between human activities, spatial patterns, and the ecological systems with which they interact. This comprehension lays the groundwork for the use of participatory GIS as a tool for looking into and evaluating these complex socio-spatial ecologies. In addition, the chapter delves into the various applications of participatory GIS in analyzing and addressing community identity and societal concerns. It goes over how PGIS can be used to investigate issues including urbanization, social disparities, environmental deterioration, and cultural landscapes. Participatory GIS allows for a better knowledge of the geographies and socio-spatial processes that define a societal area by integrating local populations in data collection, interpretation, and decision-making. The chapter also emphasizes the necessity of critical thinking and ethical considerations in participatory GIS practice. It discusses power relations, inclusion, and the representation of varied viewpoints. Therefore, this chapter provides a thorough examination of the integration of participatory GIS into sociocultural contexts. It emphasizes the need to understand the links between social, geographical, and ecological factors and highlights how participatory GIS may help to overcome complex societal concerns.

Keywords Societal challenges · Participatory GIS and community identity · Socio-spatial ecologies · People, places, and environment

In this chapter, we are going to take a fascinating look at the complex interaction between geographies and socio-spatial ecologies in societal space. Our journey begins with an introduction to participatory GIS, a powerful tool that allows communities to actively participate in mapping and understanding their surroundings. The passage discusses the exploration of participatory GIS and emphasizes its potential to democratize geographical information and foster inclusive decision-making processes. Throughout the chapter, the aim is to deepen understanding of how participatory GIS could alter perspectives on societal space and processes.

Key Points of the Chapter
- Understanding the geographies and socio-spatial ecologies of a social space.
- Investigating the spatial implications and applications of PGIS.
- Participatory spatial planning and future direction.

4.1   Understanding Societal Spaces

Societal spaces are the complex, dynamic, and interrelated environments where human civilizations live and interact. Societal space is a core notion in social geography, a branch of geography that investigates how human societies interact with and shape their physical surroundings. In social geography, societal space is the spatial dimension of human society in which social, cultural, economic, and political processes take place within a geographical setting. The following are major features of societal space in geography and spatial contexts:

- *Space as a social construct*: Social geography emphasizes that space is a social construct rather than a physical substance. It contends that space is not neutral but rather embedded with social meanings, values, and power dynamics. Social influences influence how people perceive, use, and govern their environments.
- *Urban and rural spaces*: Social geography frequently differs between urban and rural environments, delving into the distinct social and physical dynamics of each. It investigates urbanization dynamics, sprawl, and rural-urban relations.
- *Environmental impacts*: Social geography takes into account the environmental components of societal space, such as how human activities affect the natural environment and the effects of environmental changes on social systems.
- *Place and identity*: Societal space is inextricably linked to the concept of place. People form a sense of identification and attachment to certain locations, which can have a significant impact on their social interactions and behaviours. Place identity can be influenced by cultural, ethnic, or social roots.
- *Territoriality and power*: Territoriality implies the assertion of control and ownership over space. Social geography investigates how individuals, groups, and nations assert territorial rights, which frequently entails the use of power and influence.

- *Place-based interventions*: Social geography supports policy and planning interventions by emphasizing the significance of understanding sociocultural spaces in resolving challenges like urban planning, community development, and social justice.
- *Spatial patterns of inequality*: Socioeconomic geography investigates how sociocultural areas reflect and reinforce socioeconomic inequities. It investigates residential segregation, spatial differences in access to resources and opportunities, and how these spatial patterns affect various social groupings.
- *Spatial behavior and interaction*: Social geography is the study of how people and groups move through and interact with sociocultural settings. It investigates mobility, transportation, migration, and the spatial distribution of social activities.
- *Cultural landscapes*: Cultural landscapes are observable representations of human culture in the physical environment. Social geography is the study of how cultural practices such as buildings, art, and religious symbols modify and influence society.
- *Globalization and space*: In this age of globalization, social geography studies how societal places are interconnected on a global scale. It investigates the dynamics of transnational movements of people, goods, information, and ideas, as well as their implications for local and regional spaces.

In conclusion, societal space in social geography and spatial contexts is a complex and dynamic notion that acknowledges the interaction of social processes and spatial dimensions. It reveals important insights into how human societies are organized, interact, and change within their geographical surroundings. When we discuss GISciences and PGIS, we are referring to valuable instruments for analyzing and comprehending these sociocultural environments. GISciences primarily comprises the study of spatial data, mapping, and analysis (Fig. 4.1). The following are the components of this conjunction:

- Spatial data collection entails gathering geographic information via surveys, remote sensing, GPS, and other approaches.
- Data storage and management involve organizing spatial data in databases and geographic information systems.
- Geospatial analysis is the examination of geographical relationships, patterns, and trends using a variety of methods and approaches.

Key conjunction elements of PGIS are as follows:

- Community involvement: Including local information and viewpoints in the mapping process.
- Empowerment: Allowing communities the ability to handle their own spatial concerns and make educated decisions.
- Social justice: PGIS frequently focuses on underrepresented populations to promote equity and inclusion.

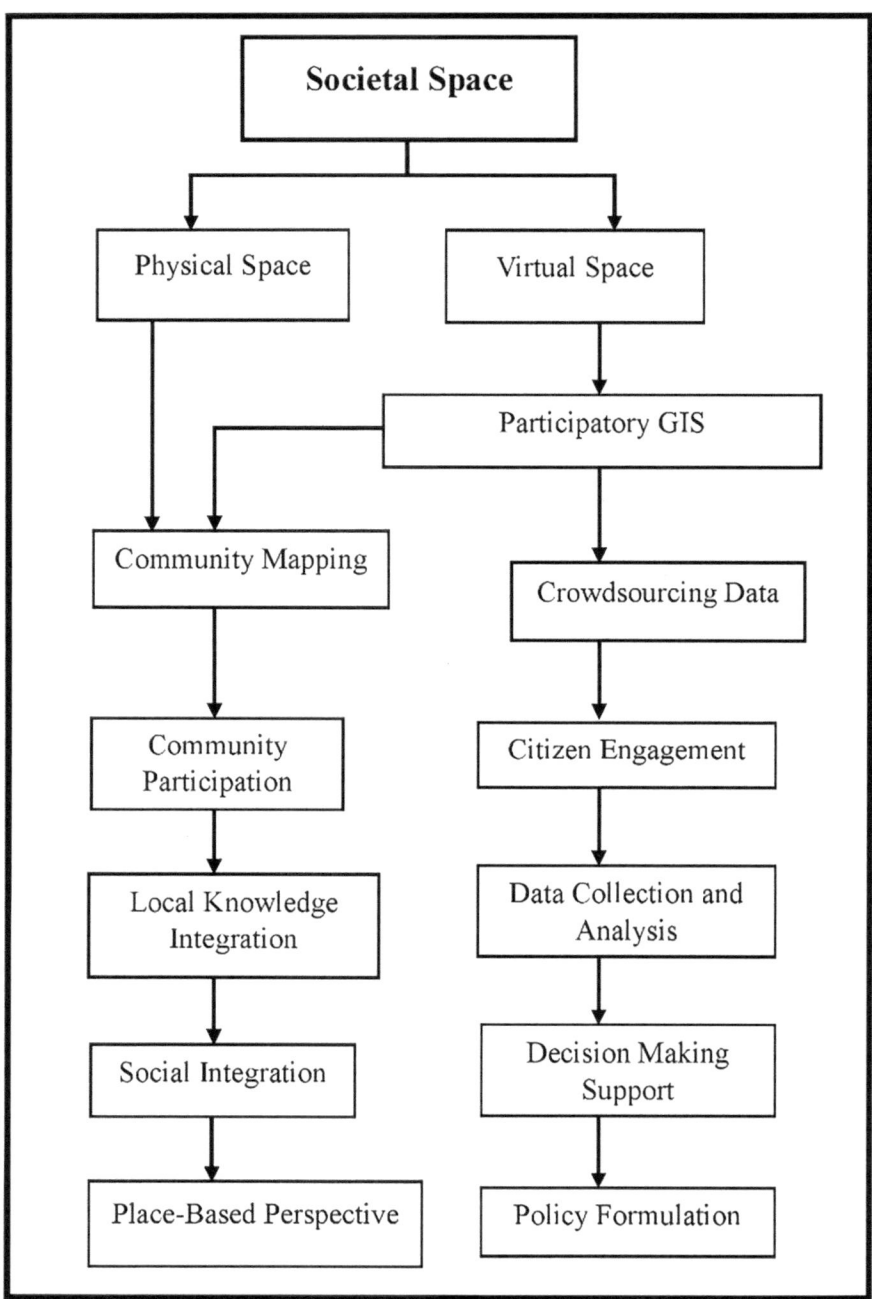

Fig. 4.1 Societal space in terms of participatory GIS. (Made by author(s))

4.2   Geographies, Socio-spatial Ecologies, and Participatory GIS

Geographies, socio-spatial ecologies, and participatory GIS are all interconnected concepts in the subject of geography and spatial analysis (Fig. 4.2). Each of these components is critical for understanding and tackling a variety of social and environmental challenges. The following explains these topics and investigates their relationships:

Geographies is the study of the Earth's physical characteristics, resources, populations, and landscapes. It is a vast field that includes both physical geography (the study of natural features) and human geography (the study of human actions and their impact on the environment). Geographers employ a variety of tools and techniques to analyze and depict geographical data, such as maps, satellite photography, and GIS. And the socio-spatial ecologies is the study of how social and environmental elements interact and impact one another within a specific geographic area. This notion highlights the interdependence of human actions and their effects on the natural environment. This field's researchers investigate subjects such as urbanization, land use patterns, environmental justice, and how communities adapt and adjust their surroundings. Incorporation of the participatory GIS approach involves local communities and stakeholders in the collection, management, and use of geographic information. It attempts to give individuals a say in decisions that affect their lives and the environment. Therefore, the specific links between these ideas are as follows:

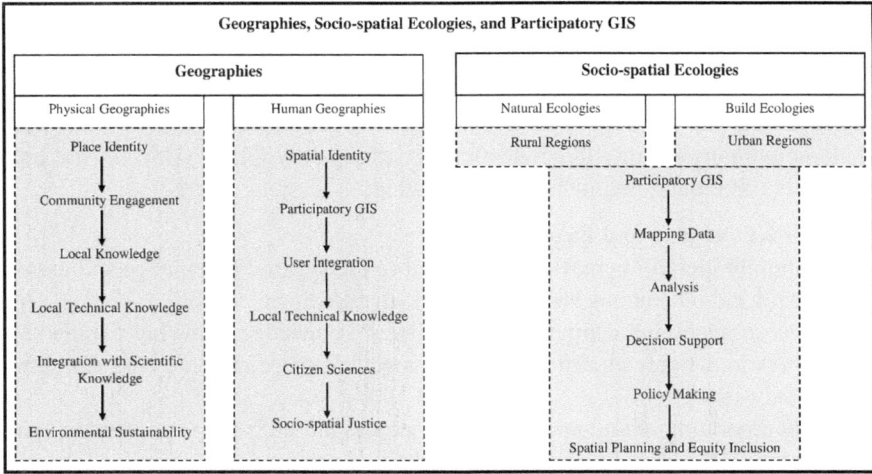

Fig. 4.2 Connection of geographies, socio-spatial ecologies, and participatory GIS. (Made by author(s))

GIS as a Tool for Geographies and Socio-spatial Ecologies
- Geographers and socio-spatial ecologists utilize GIS technology to study and visualize spatial data.
- In geography, GIS is used to map and study natural features as well as human activity on the Earth's surface.
- In socio-spatial ecology, GIS is critical for measuring the environmental impact of human activity, finding patterns, and influencing policy and planning.

Participatory GIS as a Bridge
- Participatory GIS helps bridge the gap between geography and socio-spatial ecology by including communities in geographical data collection and interpretation.
- It is especially beneficial in socio-spatial ecologies because it helps local communities understand and address the social and environmental issues that they confront.
- PGIS enables the incorporation of local knowledge and community viewpoints into larger geographic and socio-spatial analyses.

In conclusion, geographies provide the theoretical and methodological underpinnings for comprehending the Earth's spatial processes. Socio-spatial ecologies focus on the interconnections between society and the environment within a geographical space. Participatory GIS is a technique that improves both geographies and socio-spatial ecologies by including communities in spatial data collection and decision-making, making it an important tool for creating sustainable and equitable spatial settings. These ideas combine to provide a comprehensive understanding of complicated socio-environmental concerns in specific geographic situations.

4.3 Participatory GIS and Community Identity

Community identity is the common sense of belonging, values, culture, and history that unites a group of people who live in a specific geographic location. When analyzing community identity in connection to participatory GIS (PGIS), several crucial points come into play, such as the following:

(a) Local Knowledge and Expertise
- Community identity is inextricably linked to the specific knowledge and expertise that local inhabitants have about their surroundings.
- PGIS recognizes and capitalizes on this local knowledge, allowing community members to actively contribute their perspectives to geographic data gathering and analysis.
- By incorporating communities in the process, PGIS helps to protect and strengthen their identity by recognizing their perspectives and knowledge.

(b) Spatial Representation
- Community identity frequently includes a sense of place, in which certain geographic sites are culturally or historically significant.
- PGIS allows communities to map and portray these important locations, increasing their sense of identity and connection to the land.
- These spatial representations are a visual expression of community identity that can be used for campaigning, cultural preservation, and decision-making.

(c) Empowerment and Engagement
- PGIS enables communities by providing tools and technologies for active participation in geographical data gathering and analysis.
- When communities participate in environmental decision-making processes, they develop a stronger sense of identity.
- Empowered communities are more likely to take control of their surroundings and fight for changes that reflect their values and identities.

(d) Cultural Heritage Preservation
- Many communities have culturally significant sites, traditions, and behaviours that are important to their identities.
- PGIS can be used to map and document heritage resources, thereby preserving them for future generations.
- It enables communities to actively safeguard and maintain their cultural legacy, reinforcing their sense of self and connection to the past.

(e) Equitable Decision-Making
- Involving communities in GIS procedures indicates that their perspectives are heard in choices that affect their environment.
- This participatory approach encourages more egalitarian and inclusive decision-making that is consistent with the values and identity of the community.
- It helps to prevent the marginalization of specific groups and promotes social fairness in the community.

Finally, community identity and participatory GIS have a close relationship. PGIS promotes local knowledge, protects cultural heritage, and allows communities to actively engage in spatial data collection and decision-making. By incorporating community identity into GIS procedures, PGIS fosters a better connection between people and their environment, resulting in more sustainable and culturally sensitive spatial planning and development practices.

4.4 Empowering Local Communities Through Participatory GIS

Participatory PGIS is an effective technique that empowers local communities by including them in the collection, management, and use of geographic information. To further understand how PGIS strengthens local communities, consider Fig. 4.3 and the following crucial points:

Fig. 4.3 Empowering local communities through participatory GIS. (Made by author(s))

Inclusive Decision-Making
- PGIS ensures that communities participate in choices that affect their lives and ecological systems.
- It promotes inclusivity by involving marginalized or underrepresented community members in mapping and analyzing spatial data.
- Inclusive decision-making allows communities to determine their own destinies and solve issues that are important to them.

Local Knowledge Valuation
- PGIS emphasizes the importance of local knowledge and skills.
- Communities frequently have distinct perspectives on their surroundings, which include traditional practices, ecological knowledge, and cultural history.
- PGIS fosters the incorporation of local knowledge into spatial data, which improves the quality and usefulness of information.

Community Capacity Building
- Local communities benefit from PGIS by learning crucial data gathering, mapping, and spatial analysis skills.
- This capacity-building training enables community members to actively manage their resources and settings.
- It also improves their ability to advocate for their demands and communicate with government institutions and nongovernmental organizations (NGOs).

Resource Management and Environmental Stewardship
- PGIS helps communities monitor and manage their natural resources more effectively.
- Communities can record land use patterns, monitor changes in ecosystems, and identify areas of concern, resulting in more sustainable resource management methods.
- Empowered communities are better able to maintain and conserve their environment, which promotes long-term sustainability.

Social Equity and Justice
- PGIS promotes social fairness by minimizing information inequities.
- It helps to ensure that all community members have access to crucial spatial information, which is required for making educated decisions.
- This approach advances social justice by addressing imbalances in resource allocation and land tenure.

Crisis Response and Resilience
- In the event of a natural disaster or a crisis, PGIS can be an invaluable tool for quick reaction and recovery.
- Local communities can utilize PGIS to map damaged areas, identify vulnerable individuals, and coordinate emergency response efforts.
- This helps communities become more resilient in the face of adversity.

Advocacy and Policy Influence
- PGIS data can be utilized to support advocacy activities and policy discussions.
- Communities can support their requests and concerns by presenting well-documented spatial data to government officials, NGOs, and international organizations.
- This provides communities with the ability to influence policies that directly affect their lives.

In the end, participatory GIS empowers local communities by allowing them to participate in data collection and decision-making. This improves community resilience and resource management, as well as ownership, self-determination, and social justice, all of which contribute to sustainable and equitable development.

4.5 Future Directions in Socio-spatial Ecologies and Participatory GIS, Participatory Spatial Planning

The future of socio-spatial ecologies and participatory geographic information systems (GIS) will be influenced by evolving technology, shifting societal objectives, and an increasing emphasis on community engagement in addressing complex social and environmental concerns. Here are a few possible future directions for these fields:

Integration of Emerging Technologies
- In participatory GIS and socio-spatial ecologies, the usage of cutting-edge technology such as artificial intelligence, machine learning, and remote sensing will increase.
- Drone technology combined with high-resolution imagery from satellites will improve mapping and data collection capabilities.
- Large datasets may be more easily mined for insightful information by using AI and machine learning to automate tasks like pattern detection and data analysis.

Enhanced Data Accessibility and Open Data
- The movement to increase the openness and accessibility of spatial data will only intensify.
- Transparency, cooperation, and innovation will be encouraged by governments, organizations, and communities sharing their spatial data more and more.
- In order to democratize access to GIS technology, open-source GIS tools and applications will be essential.

Community-Centered Approaches
- The focus on empowerment and community involvement will only get stronger.
- With time, participatory GIS will expand to incorporate a wider range of perspectives, especially those from disadvantaged and underprivileged communities.
- One major focus will continue to be giving communities the tools they need to actively participate in data gathering, analysis, and decision-making.

Environmental Sustainability and Resilience
- The mitigation and adaptation of climate change are among the urgent environmental issues that socio-spatial ecologies will increasingly concentrate on.
- Modeling and assessing the effects of climate change on ecosystems and communities will heavily rely on GIS.
- In response to environmental risks, communities will use participatory GIS to devise solutions for boosting sustainability and resilience.

Urban Planning and Smart Cities
- GIS and participatory methods will be essential for efficient urban planning and the creation of smart cities in quickly urbanizing areas.
- In order to build more habitable and sustainable urban settings, these technologies will help optimize resource management, infrastructure, land use, and transit networks.

Health and Public Health Applications
- GIS will be useful in public health for a long time to come, with uses ranging from tracking diseases to allocating resources for treatment.
- By identifying locations with restricted access to healthcare and pushing for improvements, participatory GIS can assist communities in addressing health disparities.

Education and Capacity Building
- Particularly in nations with lower-middle incomes and others developing countries, there will be an increasing demand for GIS education and capacity building, as well as participatory techniques.
- The main goal of the initiatives is to give people and communities the tools they need to use GIS tools efficiently.

Global Collaboration and Data Sharing
- International collaboration and data sharing will be critical in solving global issues including disaster response, disease management, and sustainable development.
- Organizations and governments will collaborate to develop standards for data sharing and interoperability.

4.6 Conclusion and Arguments

The journey into participatory GIS within the context of geographies and socio-spatial ecologies demonstrates that these interconnected fields are set for an exciting and revolutionary future. The incorporation of emerging technology, a commitment to community engagement, and an emphasis on sustainability and resilience are all motivating factors. As we progress, the capacity of GIS to empower local people and bridge the gap between civilization and the environment becomes more apparent. It is a path toward inclusive decision-making, equitable resource management, and the development of smarter, more resilient cities, all of which promise a brighter, more sustainable future for our societal spaces.

Finally, the combination of geographies, socio-spatial ecologies, and participatory GIS has enormous potential for tackling the pressing challenges posed by the climate disaster, socio-ecological disturbances, and sustainability. These vital and innovative initiatives enable communities to actively shape their environments, promote resilience, and strive toward a more sustainable and equitable future.

Further Reading

Bearman, N., Jones, N., André, I., Cachinho, H. A., & DeMers, M. (2016). The future role of GIS education in creating critical spatial thinkers. *Journal of Geography in Higher Education, 40*(3), 394–408. https://doi.org/10.1080/03098265.2016.1144729

Bednarz, R. S., & Bednarz, S. W. (2008). The importance of spatial thinking in an uncertain world. In D. Z. Sui (Ed.), *Geospatial technologies and homeland security*. The GeoJournal Library, Springer. https://doi.org/10.1007/978-1-4020-8507-9_16

Buttimer, A. (1969). Social space in interdisciplinary perspective. *The Geographical Review, LIX*(3), 417–426.

Chambers, K., Corbett, J., Keller, P., & Wood, C. (2004). Indigenous knowledge, mapping, and GIS: A diffusion of innovation perspective. *Cartographica: The International Journal for Geographic Information and Geovisualization, 39*(3), 19–31. https://doi.org/10.3138/N752-N693-180T-N843

Dwivedi, A. (2023) (Ed.). *Waste management, sanitation & society*. Cambridge Scholars Publishing. ISBN: 978-1-5275-1782-0. https://www.cambridgescholars.com/product/978-1-5275-1782-0

Goodchild, M. F., & Janelle, D. G. (2010). Toward critical spatial thinking in the social sciences and humanities. *GeoJournal, 75*, 3–13.

Goodchild, M. F. (2000). The current status of GIS and spatial analysis. *Journal of Geographical Systems, 2*, 5–10.

Harrison, C., & Haklay, M. (2010). The potential of public participation geographic information systems in UK environmental planning: Appraisals by active publics. *Journal of Environmental Planning and Management, 45*(6), 841–863. https://doi.org/10.1080/0964056022000024370

Huck, J. J., Whyatt, J. D., & Coulton, P. (2014). Spraycan: A PPGIS for capturing imprecise notions of place. *Applied Geography, 55*, 229–237. https://doi.org/10.1016/j.apgeog.2014.09.007

Kyem, P. A. K. (2004). Of intractable conflicts and participatory GIS applications: The search for consensus amidst competing claims and institutional demands. *Annals of the Association of American Geographers, 94*(1), 37–57. https://doi.org/10.1111/j.1467-8306.2004.09401003.x

McCall, M. K. (2003). Seeking good governance in participatory-GIS: A review of processes and governance dimensions in applying GIS to participatory spatial planning. *Habitat International, 27*, 549–573.

Metoyer, S. K., Bednarz, S. W., & Bednarz, R. S. (2015). Spatial thinking in education: Concepts, development, and assessment. In *Geospatial technologies and geography education in a changing world* (*Advances in geographical and environmental sciences*). Springer. https://doi.org/10.1007/978-4-431-55519-3_3

Sharma, S., & Kaushik, S. P. (2016). Public participation in environment management and planning for implementing hydropower projects in the upper Beas basin, northwestern Himalaya. *The Geographers, 63*(1), 93–102.

Solari, O. M., Demirci, A., & Van Der Schee, J. A. (2015). *Geospatial technologies and geography education in a changing world* (Advances in Geographical and Environmental Sciences). Springer Nature. https://doi.org/10.1007/978-4-431-55519-3. ISBN: 978-4-431-55519-3.

Chapter 5
Community Cartography and Participatory GIS

> *Community cartography is a testament to the power of local knowledge and the importance of putting the tools of mapping in the hands of the people.*
>
> Sarah Williams

Abstract This chapter delves further into the mutually beneficial interaction between community cartography and participatory PGIS. It digs into the collaborative mapping and data collection process, emphasizing the importance of community engagement and empowerment in generating local knowledge and spatial representations. The chapter begins by describing community cartography as a participatory method of mapmaking in which local communities participate in the development and interpretation of geographical data. It highlights the importance of including local knowledge, viewpoints, and cultural context in the mapping process. Following that, the author looked at the essential components and methodologies of community cartography and participatory GIS. Explore methods such as participatory mapping, in which community members actively participate in the construction of maps, and sketch mapping, in which hand-drawn maps are used to convey local spatial knowledge. The chapter also emphasizes the use of digital technologies, such as GPS devices and mobile mapping apps, to collect and distribute geospatial data in an interactive manner. The chapter additionally looks at the advantages and applications of community cartography and participatory GIS. It addresses how these ideas might be used in a variety of contexts, including community-based resource management, urban planning, and others. Through participatory mapping procedures, the author emphasizes the importance of community ownership and empowerment, as well as the possibility of encouraging conversation, shared decision-making, and social transformation. The chapter also discusses the difficulties and concerns that come with community mapping and participatory GIS. It goes over topics like power dynamics, data quality, and the necessity for capacity growth. In addition, emphasize the importance of comprehensive data validation and integration in order to ensure the credibility and usability of community-generated spatial data. In conclusion, the chapter "Community Cartography and Participatory GIS" explores the integration of community cartography and participatory GIS in depth. It emphasizes the need to include communities in the mapping process, expand local knowledge, and promote inclusive decision-making. The chapter highlights the potential benefits, obstacles, and issues connected with these approaches, as well as the transformative power of participatory mapping in achieving community empowerment and spatial justice.

Keywords Community cartography · Participatory mapping techniques · Local knowledge and planning · Community empowerment · Spatial justice

In this chapter, we look at the transformative power of community mapping and participatory GIS. Here, we look at how communities and individuals can actively map and analyze their own places, enhancing traditional GIS with local knowledge and viewpoints. Further, this chapter talks about how spatial data is being democratized, what tools and techniques are being used, and how participatory approaches are having a significant impact on decision-making, resource management, and social justice.

Key Points of the Chapter
- Fundamentals of community cartography.
- Participatory techniques for community cartography.
- Innovations in community cartography.

5.1 Concepts and Approaches of Community Cartography

Concept
Community cartography is a multidimensional idea that focuses on collaborative mapping activities within a particular community. Community cartography is used to map informal settlements, indigenous land claims, cultural heritage locations, and community-based natural resource management. Here are some important points on the concepts of community cartography:

- *Local context*: Community cartography captures the community's distinctive features, landmarks, and challenges. This ensures that maps are appropriate to the local context and issues.
- *Community empowerment*: Community cartography enables residents to actively engage in the construction and management of maps. It uses local knowledge and viewpoints to make maps more accurate and relevant to the community's needs.
- *Bottom-up approach*: It takes a bottom-up approach, with community members being the key contributors to mapmaking. This differs from traditional cartography, which frequently relies on top-down data sources.
- *Digital tools*: Technological advancements have made community cartography more accessible by utilizing open-source mapping tools, crowdsourced data collection, and mobile apps.

- *Social and environmental impact*: Community cartography can have a good impact on both society and the environment. It helps with resource management, disaster preparedness, urban planning, and social justice efforts.
- *Community engagement*: To be sustainable, community cartography projects must engage the community actively, create trust, and collaborate on an ongoing basis.
- *Spatial literacy*: It develops spatial literacy in the community, giving people the ability to interpret and use spatial information for their own benefit.
- *Challenges*: During the mapping process, there are challenges such as guaranteeing data veracity, addressing privacy concerns, and handling community tensions.

Thus, community cartography represents a shift in the approach to map construction and application, emphasizing community engagement, local insights, and the democratization of geographic information in order to improve decision-making and community well-being.

Approaches
Community cartography uses a variety of methods to engage communities in mapmaking and geographical data collection. Here are some significant considerations about the various approaches:

- *Participatory mapping*: In this approach, community members actively draw or annotate maps to convey their local knowledge. It frequently involves sketches, notes on existing maps, and the use of participatory tools such as PPGIS (participatory geographic information systems) to build collaborative maps.
- *Crowdsourced mapping*: Using digital platforms and technology, crowdsourced mapping invites community members to provide data and information online. Users can upload and alter geographic data on platforms such as OpenStreetMap, which fosters a global mapper community.
- *Community-based GPS data collection*: Residents utilize GPS devices or smartphone apps to gather spatial data about specific features, resources, or challenges in their area. This information can then be included in maps.
- *Photovoice*: This approach mixes photography with mapping. Community members photograph key sites, events, or issues and then utilize the images to construct maps that transmit both geographical and visual data.
- *Storytelling maps*: These maps combine narrative and spatial data, allowing community members to express their stories and experiences using maps. These maps may have audio, video, or written narratives related to certain locations.
- *Community workshops and trainings*: Training sessions and workshops are held to educate community members about mapping tools and procedures. These events provide individuals with the necessary skills to actively participate in mapmaking.
- *Participatory GIS*: PGIS integrates geospatial information systems with participatory techniques. It engages community participants in all aspects of GIS initiatives, from data collection to analysis and decision-making.

- *Community surveys*: Structured surveys are done within communities to collect spatial data on a variety of topics, including land use, infrastructure demands, and environmental concerns. The survey results are subsequently translated into maps.
- *Community asset mapping*: This approach identifies and maps community assets such as physical resources, cultural legacy, and social networks. It enables communities to understand their own strengths and resources.
- *Public gatherings and deliberative mapping*: Community gatherings or forums are convened to address spatial concerns and make collaborative decisions. Deliberative mapping strategies help groups make decisions and reach consensus.
- *Interactive web maps*: Community-generated maps can be made available online, allowing residents and outside stakeholders to engage with and explore geographical data. This encourages transparency and increased participation.
- *Collaborative data platforms*: Online platforms and tools like Google My Maps and ArcGIS Online allow community members to create and share maps together.

Each of these approaches may be tailored to a community's individual requirements and skills, making community cartography a versatile and inclusive methodology for community engagement in the mapping process.

5.2   Participatory Mapping Techniques for Community Cartography

Participatory mapping approaches are central to community cartography, allowing for active community engagement in map construction. These strategies allow communities to express their local knowledge, goals, and concerns. Here's an overview of the numerous participatory mapping approaches used in community cartography:

- *Sketch maps*: Sketch maps are simple and accessible, with community members drawing maps by hand. They may depict local features, resources, or issues. Sketch maps are often used as a starting point for more detailed mapping projects.
- *Community walks*: Residents stroll across their neighborhoods, noting noteworthy features, risks, or points of interest on a physical map or digital gadget. This approach combines spatial data collection and on-the-ground observations.
- *Participatory GPS mapping*: Community participants collect precise spatial data using GPS devices or smartphone apps. This information can then be combined into geographic information systems (GIS) to generate reliable community maps.
- *Transect mapping*: Transect mapping is drawing a straight line through a community and documenting observations along it. This approach creates a cross-sectional image of the area, emphasizing spatial differences and characteristics.
- *Village resource mapping*: This method involves identifying and mapping community resources such as water sources, schools, health facilities, and markets. It contributes to a better understanding of how important services are distributed.

- *Historical mapping*: By using maps, community members can document the historical evolution of their area. This includes changes in land use, infrastructure, and demography over time.
- *Photomapping*: By combining photography and mapping, community members photograph and geotag notable areas or issues. The final map contains images that give spatial information and context.
- *3D mapping*: In some circumstances, communities may employ 3D modeling technologies to generate maps that depict terrain and topography, which can benefit disaster preparedness or urban planning.
- *Participatory GIS*: PGIS incorporates GIS technology and involves community participants in all stages of GIS initiatives, such as data gathering, analysis, and decision-making. It allows communities to efficiently access and use spatial data.
- *Storytelling maps*: Using maps, community members offer stories about specific locales. These stories can take the form of audio, video, or written narratives, bringing a human touch to the spatial data.
- *Public meetings and deliberative mapping*: Community meetings or workshops are held to discuss and map spatial issues. Deliberative mapping methods, such as multi-criteria analysis, aid in collective decision-making and prioritization.
- *Interactive web maps*: Community-generated maps can be made available online, allowing residents and outside stakeholders to interact with the spatial data. This encourages transparency and increased participation.
- *Collaborative data platforms*: Online platforms and tools, such as Google My Maps and ArcGIS Online, allow community members to create and share maps and data from a distance.
- *Asset mapping*: This technique identifies communal assets and helps communities identify their strengths, resources, and development potential. It may contain physical, social, and cultural assets.
- *Temporal mapping*: Temporal mapping is a technique for recording changes in a specific area across time. It is useful for tracking environmental changes, land use shifts, and the effects of development projects.

To summarize, participatory mapping techniques allow communities to become active participants in the creation of maps that reflect their own viewpoints and needs. These strategies promote community empowerment, data ownership, and informed decision-making, making them indispensable tools for community cartography.

5.3 Empowering Communities Through Participatory GIS

Empowering communities through participatory GIS (PGIS) is a collaborative strategy that provides communities with the tools and expertise they need to gather, analyze, and apply geospatial data. By incorporating local knowledge and viewpoints into geographical data, PGIS ensures that maps and information appropriately represent the community's needs and goals. This empowerment encompasses decision-making procedures, resource management, emergency preparedness, and social justice activities. PGIS promotes transparency, accountability, and data ownership, hence increasing community members' digital literacy and ability to solve their own challenges and possibilities. It encourages collaboration, sustainable development, and the protection of cultural heritage while empowering communities to take control of their geographical data and capitalize on its potential for good change. Participatory geographic information systems (PGIS) is an effective technique that uses GIS technology to empower communities in a variety of ways such as the following:

- *Local knowledge integration*: PGIS enables communities to incorporate their own local knowledge and viewpoints into spatial data. This guarantees that maps and spatial data are culturally and contextually appropriate.
- *Community engagement*: It encourages active community participation in data collection, analysis, and decision-making connected to geographic information. This involvement enables people to take control of their spatial data.
- *Enhanced decision-making*: Using PGIS, communities may make informed decisions based on geographical data. It assists in prioritizing resource allocation, planning development initiatives, and addressing critical concerns such as disaster management and land use planning.
- *Resource management*: PGIS can help communities manage and sustain natural resources like forests, water bodies, and agricultural land. This can result in better livelihoods and environmental conservation.
- *Social justice and advocacy*: PGIS can help with social justice projects by mapping issues such as land tenure, indigenous areas, and marginalized communities. It provides useful evidence for advocacy and legal action.
- *Disaster preparedness*: PGIS can help communities map and assess natural disaster vulnerabilities, as well as organize disaster preparedness and response actions. This can help save lives and prevent damage during an emergency.
- *Education and capacity building*: PGIS programs frequently entail teaching community members about GIS technology and data collection techniques. This improves their digital literacy and analysis skills.
- *Transparency and accountability*: By making spatial data available to community members and external stakeholders, PGIS enhances transparency in governance and decision-making processes.
- *Cultural preservation*: PGIS can help to document and preserve cultural heritage sites and traditions, ensuring that they are recognized and safeguarded.
- *Collaboration and networking*: PGIS promotes collaboration between communities, organizations, and experts. It promotes the exchange of knowledge, best practices, and resources.
- *Data ownership*: In PGIS efforts, communities retain sovereignty over their own data, decreasing the danger of exploitation and guaranteeing that information is used to benefit the community.

- *Sustainable development*: PGIS helps communities plan infrastructure projects, monitor land use changes, and assess the environmental impact of development operations.

In brief, participatory GIS empowers communities by providing the tools and knowledge they need to actively manage local geospatial information. It promotes a sense of ownership, enhances decision-making, and aids in community development and well-being.

5.4 Participatory GIS for Engaging Local Knowledge, Community Development, and Planning in Community Cartography

Participatory GIS serve an important role in engaging local knowledge, supporting community development, and enabling successful planning in the field of community cartography. This comprehensive discussions investigates how PGIS accomplishes these goals and its value in empowering communities.

Engaging Local Knowledge
PGIS prioritizes local knowledge while collecting and analyzing spatial data. It acknowledges that communities have invaluable knowledge of their environment, including cultural practices, natural resources, and local challenges. By integrating community members in mapping processes, PGIS preserves, documents, and integrates local knowledge into GIS databases and maps. This method produces more precise and contextually relevant geographic information.

Community Development
Participatory GIS serves as a catalyst for community development on multiple fronts:

- *Resource management*: PGIS can help communities manage and use their natural resources more sustainably, resulting in improved livelihoods and environmental conservation.
- *Infrastructure planning*: PGIS enables communities to plan infrastructure projects such as schools, health clinics, and roads based on their needs and priorities, so supporting development that is consistent with their aspirations.
- *Disaster preparedness*: Using PGIS, communities may map their vulnerability to natural disasters, improving their ability to prepare for and respond to catastrophes, eventually saving lives and minimizing damage.
- *Cultural preservation*: PGIS helps to document and preserve cultural heritage sites and traditions, ensuring that they are identified, preserved, and passed down to future generations.

Planning for Community Cartography
In the framework of community cartography, PGIS improves planning procedures in various ways.

- *Participatory land use planning*: PGIS promotes land use planning that takes into account local practices and community preferences, hence decreasing conflicts over land and resources.
- *Social justice initiatives*: By mapping issues such as land tenure, indigenous territory, and marginalized groups, PGIS serves as evidence for advocacy and legal activities that promote social justice and equitable resource distribution.
- *Transparency and responsibility*: PGIS improves transparency in governance by making spatial data available to community members and external stakeholders, hence increasing responsibility in decision-making processes.

In short, participatory GIS is an essential tool for engaging local knowledge, supporting community development, and enabling successful planning in the field of community cartography. It enables communities to take control of their spatial data, promotes informed and participatory decision-making, and contributes to holistic development that is consistent with community ambitions and values.

5.5   Challenges, Innovations, and Future Directions in Community Cartography

Community cartography, which involves local communities in the construction and use of maps, is a rapidly evolving field with both obstacles and exciting future paths. Here's an in-depth discussion of these topics as well as advances in the field:

Challenges
- *Digital divide*: Access to technology and digital literacy can be impediments to participation, restricting the scope of community cartography programs.
- *Data privacy and ethics*: Protecting sensitive information and guaranteeing the ethical use of community data is an ongoing problem, especially in scenarios with vulnerable populations.
- *Data accuracy and quality*: Community-contributed data might vary in accuracy and reliability, raising questions about the quality of community-generated maps.
- *Community engagement*: Maintaining community interest and participation in mapping activities over time can be difficult, necessitating continuing efforts in capacity building and motivation.
- *Conflict resolution*: Mapping can occasionally uncover disputes within communities, which must be handled cautiously to avoid aggravating tensions.
- *Standardization*: Ensuring that community-generated data can be properly merged with official datasets frequently necessitates standardization initiatives and collaboration with government entities.
- *Sustainability*: Many community mapping projects struggle to maintain momentum and funding after their initial phases, which has an influence on data and engagement.

Innovations
- *Community land titling*: Using mapping to promote land tenure and titling operations can empower communities while also protecting their land rights.
- *Mixed reality mapping*: It uses augmented and virtual reality technologies to build interactive and immersive community maps.
- *Storytelling maps*: Combining maps with tales, photographs, and videos improves the storytelling component of community mapping, making it more engaging and instructive.
- *Community-led drones and UAVs*: Communities are increasingly using drones to collect aerial data for mapping and disaster response.
- *Community-driven sensor networks*: Communities are establishing sensor networks to collect real-time data on environmental conditions, pollution, and other variables.
- *Participatory 3D mapping*: Using 3D modeling techniques, communities can build more comprehensive and immersive maps.

Future Directions
- *Community-centric platforms*: Customized platforms created for community use can improve data collection and map production while also increasing community engagement.
- *Advanced data collection tools*: Advancements in mobile technology and data collection apps make it easier for communities to collect and share spatial data.
- *Open data initiatives*: Providing open access to community-generated data can lead to greater use and effect.
- *Machine learning and artificial intelligence*: These technologies can help automate the study of community-generated data, increasing its accuracy and usefulness.
- *Integration with official systems*: Collaborating with government agencies and incorporating community data into official systems can improve data dependability and utility.

5.6   Socio-spatial Justice and Community Cartography

Community cartography plays an important role in promoting social and spatial justice by empowering communities to address gaps and inequities in their local contexts. The following discussion helps with this sort of justice:

5.6.1   Spatial Justice

- *Land use planning*: Participatory mapping gives communities a voice in land use planning. They can make sure that urban development and zoning rules do not discriminate against specific neighborhoods or groups.

- *Resource distribution*: Community mapping assists communities in identifying inequities in resource distribution, including infrastructure, healthcare facilities, and educational institutions. Mapping these gaps allows communities to argue for more equitable distribution.
- *Transportation equity*: Communities can map public transportation networks to determine whether they serve all areas equally. This can lead to enhanced transit access for underserved neighborhoods.
- *Environmental justice*: Mapping pollution sources, hazardous waste sites, and disaster zones enables communities to advocate for environmental justice. They can demand cleaner air and better protection from environmental threats.

5.6.2   Social Justice

- *Disaster preparedness and response*: The identification of vulnerabilities and hazards within a community enhances fair disaster preparedness, ensuring that natural disasters do not disproportionately affect vulnerable communities.
- *Community engagement*: The participatory mapping approach promotes social justice by incorporating community members in decision-making and giving them a voice on issues that affect their lives.
- *Community organizing*: By mapping concerns such as healthcare, education, and housing, communities can mobilize and represent for better services and social justice.
- *Cultural heritage preservation*: Participatory mapping contributes to the documentation and protection of cultural heritage places, customs, and indigenous territories, fostering social justice by recognizing and protecting cultural identities.
- *Legal advocacy*: Maps developed through community cartography can be used as strong evidence in judicial proceedings involving social injustices, such as land conflicts or environmental damage.
- *Land tenure and property rights*: Mapping land boundaries and documenting land tenure can be critical in protecting marginalized people's property rights and preventing them from being unfairly relocated.

In essence, community cartography enables communities to detect and remedy spatial and social injustices by giving them the skills and knowledge they need to fight for fair resource allocation, land use, and social services. It fosters transparency, accountability, and inclusivity in decision-making processes, resulting in more just and equitable communities.

5.7 Conclusion and Arguments

In conclusion, the combination of community cartography and participatory GIS creates a dynamic sector ripe for research and invention. The scope of research in this area is enormous, spanning technological developments, ethical considerations, social empowerment, and environmental sustainability. By leveraging emerging technologies, addressing data quality and inclusivity issues, and promoting ethical practices, PGIS can continue to empower communities, drive social and spatial justice, and play an important role in addressing complex global challenges like climate change and disaster response. The future of community cartography in relation to PGIS looks promising as a catalyst for positive change and community-led solutions in an ever-changing environment. In this sense, the following convention serves as the basis for further discussions:

- *Environmental sustainability*: PGIS helps communities manage resources sustainably and adapt to climate change by mapping vulnerabilities and monitoring ecological changes.
- *Empowering communities*: Combining community cartography and participatory GIS allows local communities to actively participate in mapping their surroundings, democratizing spatial data and decision-making processes.
- *Data accuracy and trust*: While community-generated data can be extremely useful, ensuring its accuracy and dependability remains difficult. Innovations in validation methodologies and blockchain technology can help increase data trust.
- *Inclusivity*: Bridging the digital divide and including marginalized groups in PGIS projects is critical for fair representation and decision-making.
- *Privacy and ethics*: The ethical collection and use of community data is critical. Researchers need to create frameworks that stress informed consent, data ownership, and data privacy.
- *Cultural preservation*: Community cartography, in conjunction with PGIS, helps to document and preserve cultural history and indigenous knowledge, promoting cultural preservation and acknowledgment.
- *Community-based disaster response*: Rapid mapping using PGIS is an innovative approach to disaster response that allows for fast coordination and relief operations in affected communities.
- *Cross-disciplinary collaboration*: Interdisciplinary research that combines PGIS with subjects such as public health, urban planning, and ecology enables novel answers to complex societal and environmental concerns.
- *Sustainability and scalability*: Ensuring the long-term viability and scalability of PGIS programs is critical to their ongoing effectiveness and relevance.
- *Policy impact*: Community-generated maps influence policy formation and execution, highlighting PGIS's ability to create positive policy changes at many levels of government.

Further Reading

Aitken, S. C. (2002). Public participation, technological discourses and the scale of GIS. In W. J. Craig, T. M. Harris, & D. Weiner (Eds.), *Community participation and geographic information systems* (pp. 357–366). Taylor and Francis.

Barndt, M. (1998). Public participation GIS – Barriers to implementation. *Cartography and Geographic Information Systems, 25*, 105–112.

Brown, G. (2016). A review of sampling effects and response bias in internet participatory mapping (PPGIS/PGIS/VGI). *Transactions in GIS, 21*, 39. https://doi.org/10.1111/tgis.12207

Brown, G., et al. (2017). Mixed methods participatory GIS: An evaluation of the validity of qualitative and quantitative mapping methods. *Applied Geography, 79*, 153–166. https://doi.org/10.1016/J.APGEOG.2016.12.015

Brown, G., et al. (2017). Identifying environmental and natural resource management conflict potential using participatory mapping. *Society & Natural Resources, 30*(12), 1458–1475. https://doi.org/10.1080/08941920.2017.1347977

Craig, W., Harris, T., & Weiner, D. (2002). *Community participation and geographic information systems*. CRC Press (Taylor and Francis). https://doi.org/10.1201/9780203469484. ISBN: 9780203469484.

Elwood, S. (2006). Participatory GIS and community planning: Restructuring technologies, social processes, and future research in PPGIS. In *Collaborative geographic information systems* (pp. 66–84). IGI Global. https://doi.org/10.4018/978-1-59140-845-1.ch004

Engen, S., Runge, C., Brown, G., Fauchald, P., Nilsen, L., & Hausner, V. (2018). Assessing local acceptance of protected area management using public participation GIS (PPGIS). *Journal for Nature Conservation, 43*, 27–34. https://doi.org/10.1016/j.jnc.2017.12.002

King, B. H. (2002). Towards a participatory GIS: Evaluating case studies of participatory rural appraisal and GIS in the developing world. *Cartography and Geographic Information Science, 29*, 43–52.

Plantin, J. C. (2014). *Participatory mapping: New data, new cartography*. Wiley. ISBN: 978-1848216617.

Rundstrom, R. A. (1995). GIS, indigenous peoples, and epistemological diversity. *Cartography and Geographic Information Systems, 22*, 45–57.

Sharma, S. (2011). Strategies for technical sustainable development of hydropower projects in the mountain environment by adopting participatory approach. *Indian Journal of Power and River Valley Development, 61*(9), 147–153.

Chapter 6
GIS for All: Challenges and Future Directions

> *GIS technology gives a voice to the voiceless and a vision to the visionaries.*
>
> John Aubrey Douglass

Abstract The chapter "GIS for All: Challenges and Future Directions" addresses the obstacles and potential future advances in the field of geographic information systems in order to make GIS accessible and valuable to all persons and communities. It emphasizes the significance of removing obstacles, encouraging inclusivity, and leveraging the revolutionary power of GIS technology. The chapter begins with a discussion of the present issues and limits of GIS, especially PGIS, that are impeding its widespread acceptance and accessibility. It addresses challenges including cost and affordability, technical complexity, and data scarcity. In addition, the chapter discusses the significance of encouraging inclusion in GIS. It emphasizes the importance of addressing gaps in GIS technology and geographical data availability, particularly in marginalized and underserved populations. It discusses the benefits and problems of promoting PGIS literacy, open-source GIS, free tools, web-based GIS, mobile GIS, and crowd-sourced mapping. The chapter also discusses potential future prospects for GIS (including PGIS) and suggests areas where progress is required. It emphasizes the significance of user-friendly interfaces and intuitive technologies that require little technical knowledge. Furthermore, the chapter discusses emerging trends and prospects in GIS. It looks at how emerging technologies like cloud computing, mobile mapping, and artificial intelligence can improve the accessibility and capabilities of GIS. It also delves into the growing relevance of combining GIS with other disciplines, such as social sciences and public health, in order to address complex societal concerns. Finally, the chapter concludes the significance of making GIS accessible and valuable to all individuals and communities. It encourages stakeholders in academia, industry, and government to work together to address the difficulties and progress in the field of GIS. GIS can become a strong instrument for informed decision-making, sustainable development, and social fairness through fostering inclusivity, technological advancements, and responsible practices. The chapter emphasizes the importance of technological breakthroughs, data sharing, and interdisciplinary cooperation in addressing complex societal concerns and realizing the transformative potential of GIS.

Keywords PGIS literacy · Future of PGIS technology · Sustainable development and social equity · Mobile GIS · Technological advancements

In this chapter, we are going to take an exploration through the realm of geographic information systems (GIS), with a focus on inclusion and accessibility. While GIS has achieved significant advances in a variety of disciplines, achieving its full potential for all people and communities remains a problem. We address the barriers to equitable GIS use, including concerns relating to technology, data, and education. Furthermore, we look at possible future possibilities in GIS, such as the democratization of geospatial technologies and GIS' role in tackling global concerns. Join us as we develop the ever-evolving GIS landscape, striving for a society in which spatial information really benefits everyone.

Key Points of the Chapter
- Understanding the challenges and future directions of GIS.
- Exploring open-source GIS and free tools.
- Promoting PGIS literacy.

6.1 Today's GIS and Its Limitations

Today's geographic information systems (GIS) have greatly improved, yet they still have several limitations, particularly those linked with participatory GIS. The following are several important aspects:

- *Data quality and accuracy*: While GIS technology has made data more accurate, errors and inaccuracies can still occur, influencing decision-making. PGIS, in particular, may experience difficulties in ensuring data quality due to its reliance on community-generated content.
- *Scale and resolution*: GIS data's scale and resolution can vary, restricting its usefulness for specific purposes. PGIS projects may struggle to obtain high-resolution data for localized mapping.
- *Data standards and interoperability*: GIS data frequently comes in a variety of forms and standards, which might impede data sharing and interoperability. This issue is especially pronounced in PGIS, where data may not meet industry requirements.
- *Data bias*: GIS data has the potential to sustain biases if not gathered and assessed meticulously. In PGIS, community data might mirror local viewpoints but remain susceptible to social biases.
- *Data privacy and security*: GIS systems frequently process sensitive spatial data. Ensuring the privacy and security of this data, particularly in PGIS with community-contributed data, is a top priority.

- *Technical complexity*: GIS software can be difficult and require specialist training, making it unavailable to people with no technical expertise. To overcome this barrier, PGIS efforts must invest in community training programs.
- *Institutional barriers*: Bureaucratic and institutional barriers can hamper data exchange and collaboration, reducing the effectiveness of GIS efforts, such as PGIS.
- *Ethical concerns*: The ethical use of GIS data, particularly in PGIS, poses issues of informed permission, data ownership, and the possibility of data misuse or exploitation.
- *Scalability*: PGIS projects frequently start at the community level, and expanding them to include greater areas or populations while preserving community engagement can be difficult.

Despite its limitations, modern GIS, especially PGIS, has enormous potential for informed decision-making, community participation, environmental management, and social justice. Addressing these difficulties through technology innovation, ethical principles, and inclusive legislation can help individuals and communities realize the full potential of GIS.

6.2 Challenges in Democratizing PGIS

Democratizing participatory GIS presents various issues that must be carefully addressed. Bridging the digital gap is a major challenge since many communities lack access to the technology and internet connectivity needed for PGIS participation. Furthermore, the technical complexity of PGIS tools and software can be a barrier, necessitating streamlined interfaces and easy-to-use training materials. The costs of obtaining and maintaining PGIS technology, combined with the requirement to expand capacity, can put pressure on resource-constrained communities. Data quality assurance and validation present extra hurdles for verifying the accuracy of community-contributed data. Maintaining data privacy and security while adhering to local cultural norms and customs is likewise a challenging task. Furthermore, inclusion remains a top goal, necessitating initiatives to involve marginalized groups and foster trust among communities. Addressing these issues is critical to making PGIS a more inclusive and equitable tool for community interaction and decision-making.

Democratizing participatory geographic information systems (PGIS) entails making the technology and approaches available to a diverse range of individuals and groups. However, there are various hurdles to reaching this goal such as the following:

- *Digital divide*: Many communities, particularly those in rural or underdeveloped areas, may not have access to the equipment and internet connectivity needed for PGIS. Bridging the digital divide is critical to ensuring fair participation.

- *Costs*: Purchasing and maintaining PGIS technology can be expensive, especially in resource-constrained regions. Funding and resources are required to support PGIS activities.
- *Technical complexity*: PGIS software and tools can be technically complicated, necessitating training and expertise. Simplifying user interfaces and providing user-friendly guidelines are critical for increased adoption.
- *Capacity building*: Developing the skills and knowledge required for effective involvement in PGIS programs is a big undertaking. Communities frequently require training and continuous support.
- *Data ownership*: Identifying data ownership and control can be difficult, especially when several parties are involved. Clear agreements and frameworks are required to handle this situation.
- *Data integration*: Combining community-generated data with government datasets can be difficult due to differences in data formats, standards, and governance.
- *Data quality*: Maintaining the accuracy and dependability of community-contributed data in PGIS can be difficult. Data validation and quality assurance techniques are required to overcome this issue.
- *Data privacy*: PGIS involves the collection and exchange of spatial data, which raises privacy and security concerns. Creating protocols for data handling and protection is critical.
- *Community empowerment*: Ensuring that PGIS programs truly empower communities and impact decision-making processes can be difficult, as some initiatives may be symbolic rather than revolutionary.
- *Cultural sensitivity*: PGIS initiatives must adhere to local cultural norms and customs, which may vary greatly. Cultural sensitivity and community engagement are critical for fostering trust and inclusivity.
- *Inclusivity*: Ensuring that underprivileged or vulnerable groups within communities are included in PGIS programs can be difficult. Efforts must be made to engage all community members, particularly those with minimal voice or representation.
- *Scalability*: Many PGIS programs start at the community level, and scaling them up while maintaining community engagement and control can be problematic.
- *Policy and governance*: Existing rules and governance structures may conflict with PGIS principles, making it difficult to implement participatory mapping and data-sharing practices.

Addressing these obstacles to democratizing PGIS necessitates a collaborative effort among community stakeholders, policymakers, technology developers, and scholars. To ensure that the benefits of PGIS are available to all, adjustments in policy, budget allocation, and capacity-building efforts are required, in addition to technology solutions.

6.3 Promoting PGIS Literacy

Promoting PGIS literacy is an important effort that enables communities and individuals to benefit from the benefits of geospatial technology for community development, environmental management, and social justice. Following are major techniques to enhance PGIS literacy:

- *Community-based training*: Provide hands-on training sessions in communities to help people develop PGIS abilities. Customize these workshops to meet local requirements and make them accessible to all community members.
- *Community mapping initiatives*: Actively engage communities in mapping their own settings, fostering a sense of ownership and importance.
- *Youth engagement*: Engage young community members in PGIS activities, since they frequently have a natural affinity for technology and can play an important role in passing on information.
- *Local language and culture*: Provide PGIS materials and training in local languages and culturally sensitive formats to increase community members' knowledge and engagement.
- *Data validation*: Teach community members how to validate and check data so that it is more accurate and reliable.
- *Accessible technology*: Make affordable or open-source GIS software and tools available on community computers, as well as mobile devices.
- *Continuous learning*: Create systems for ongoing PGIS learning and assistance, recognizing that proficiency grows over time.
- *Peer learning*: Encourage community members to share their knowledge with one another. Experienced PGIS users can coach and assist novices.
- *Practical projects*: Implement real-world PGIS projects in the community to demonstrate the practical applications of geospatial data and maps in addressing local challenges.
- *Cross-sector collaboration*: Encourage collaboration across sectors, such as education, technology, and community development, to provide comprehensive support for PGIS literacy.
- *Collaboration with NGOs and academia*: Work with nongovernmental organizations (NGOs), universities, and research institutions to provide resources, expertise, and financing for PGIS literacy programs.
- *Digital storytelling*: Use PGIS and storytelling approaches to help communities share their experiences, cultural legacy, and local knowledge.
- *Advocacy skills*: Provide participants with the necessary skills to effectively convey their PGIS results and concerns to local authorities, policymakers, and stakeholders.
- *Customized curriculum*: Produce PGIS training materials and curriculum tailored to the community's individual needs and interests, with a focus on practical applications.
- *Ethical guidelines*: Educate participants on the ethical implications of data collection, including informed permission and appropriate data handling.

Thus, improving PGIS literacy not only strengthens a community's ability to address its specific concerns, but it also leads to more equitable, informed, and inclusive decision-making. By emphasizing accessibility, cultural sensitivity, and continuous assistance, PGIS literacy can be a strong tool for community empowerment and sustainable development.

Advantages of PGIS Literacy
- *Advocacy*: PGIS literacy strengthens advocacy efforts by enabling communities to communicate their needs, objectives, and concerns to policymakers and stakeholders.
- *Empowerment*: PGIS literacy enables individuals and communities to actively participate in environmental decision-making processes, which fosters a sense of ownership and agency.
- *Informed decision-making*: PGIS-literate persons may make informed decisions by analyzing and interpreting spatial data, leading to more effective community-driven initiatives.
- *Cross-disciplinary collaboration*: People who are familiar with PGIS can work together to solve complicated problems in sectors such as public health, urban planning, and environmental sciences.
- *Disaster preparedness*: Communities that understand PGIS are better able to identify natural disaster vulnerabilities, plan for emergencies, and respond effectively during crises.
- *Resource management*: PGIS literacy aids sustainable resource management by enabling people to monitor and manage natural resources, such as forests or water sources.
- *Cultural preservation*: PGIS can be used to document and maintain cultural heritage locations and practices, encouraging the acknowledgment and preservation of cultural identities.

Challenges in Promoting PGIS Literacy
- *Data quality*: Ensuring the correctness and dependability of community-generated data is difficult, necessitating quality assurance and validation procedures.
- *Data privacy and security*: Working with geographical data presents privacy and security concerns, demanding the creation of standards and ethical norms.
- *Digital divide*: Access to technology and digital literacy skills are unevenly distributed, with marginalized populations frequently lacking the means and training required for PGIS proficiency.
- *Technical difficulty*: PGIS tools and software can be technically hard, necessitating training and support that is not always readily available.
- *Inclusivity*: Ensuring that marginalized or vulnerable groups in communities participate in PGIS literacy initiatives necessitates specific outreach and support.
- *Cost*: Acquiring and maintaining technology for PGIS literacy can be prohibitively expensive for low-income populations.
- *Cultural sensitivity*: PGIS efforts must be culturally sensitive and adhere to local conventions and customs, which can vary significantly.

6.4   Open-Source GIS and Free Tools

Open-source GIS and free GIS tools have made geospatial technology more accessible to a wider variety of scholars. The following are some noteworthy open-source GIS software and free GIS tools:

- *QGIS (quantum GIS)*: QGIS is a popular open-source desktop GIS software. It provides an easy-to-use interface, comprehensive geographic capabilities, and a thriving user and developer community. QGIS supports a variety of data formats and plugins for customization.
- *SAGA GIS*: SAGA GIS is an open-source geoscience analysis software. It includes a full suite of tools for terrain analysis, picture processing, spatial statistics, and more. It is well-known for its versatility and ease of usage.
- *GRASS GIS*: The geographic resources analysis support system (GRASS) is a robust open-source GIS that focuses on environmental modeling, remote sensing, and geospatial analysis. It offers both a command-line and graphical user interface (GUI) options.
- *gvSIG*: gvSIG is an open-source GIS software created in Spain. It has a user-friendly interface and 3D visualization capabilities and supports a variety of data formats. It is very popular in Spanish-speaking areas.
- *uDig*: User-friendly Desktop Internet GIS (uDig) is an open-source desktop GIS application that is suitable for both beginners and experts. It offers tools for data editing, map creation, and spatial analysis.
- *MapServer*: MapServer is an open-source development platform for creating web-based GIS applications. It allows for the production of dynamic, interactive maps and geographical data services.
- *OpenLayers*: OpenLayers is an open-source JavaScript library for producing interactive maps in web applications. It enables users to view and interact with maps from a variety of data sources, including web services.
- *GeoServer*: GeoServer is an open-source server program that allows you to share, process, and change geospatial data on the web. It supports standards such as Web Map Service (WMS), Web Feature Service (WFS), and Web Coverage Service (WCS).
- *GeoDa*: GeoDa is an open-source software program used for spatial data exploration and econometric research. It is especially valuable for academics and policymakers dealing with spatial data.
- *Leaflet*: Leaflet is an open-source JavaScript library for making interactive maps for websites. It's lightweight, mobile-friendly, and simple to combine with other web technologies.
- *Whitebox GAT*: Whitebox Geospatial Analysis Tools (GAT) is an open-source geospatial software that allows for advanced spatial analysis. It features a variety of hydrology, landscape analysis, and picture processing applications.
- *PostGIS*: PostGIS is an open-source PostgreSQL geographic database extension. It has support for spatial objects, allowing users to execute advanced spatial queries and analysis in a relational database.

As a consequence, these open-source GIS software and free GIS tools offer cost-effective alternatives to proprietary solutions, making geospatial technology more accessible to researchers, developers, and organizations with limited resources. As well, these encourage innovation, collaboration, and the democratization of geographical data processing and mapping. Integrating participatory GIS with open-source GIS and free technologies is an effective way to maximize the benefits of community interaction and geospatial research.

6.5 Web-Based GIS, Mobile GIS, and Crowdsourced Mapping

6.5.1 Web-Based GIS

Web-based GIS: Web-based GIS is the use of the internet and online technologies to access, visualize, analyze, and distribute geospatial data and maps. It enables people to interact with maps and spatial information via web browsers, making geospatial data more accessible and collaborative.

Relationship Between Web-Based GIS and PGIS
- Community engagement: PGIS frequently engages community members in data collection and mapping. Web-based GIS platforms can increase involvement by providing web interfaces for community members to input data and view maps online.
- Real-time data: Web-based GIS can incorporate real-time data feeds, which might be useful for PGIS projects involving disaster response, environmental monitoring, or other dynamic scenarios.
- Data sharing: Web-based GIS enables the easy sharing of maps and geographic data with a larger audience, including policymakers and the general public. This accords with PGIS's mission of raising awareness and influencing decision-makers.
- Accessibility: Web-based GIS improves the availability of geospatial data and tools, which is consistent with the participatory character of PGIS. It enables communities to view and share geospatial data from anywhere with an internet connection.
- Collaboration: Collaboration and data sharing are key features of both web-based GIS and PGIS. Web-based tools make collaborative mapping easier, allowing various stakeholders to collaborate on geographical initiatives.
- Mapping apps: Web-based GIS platforms frequently provide configurable mapping apps that can be adjusted to the specific requirements of PGIS projects, enhancing community interaction and data presentation.

In short, Web-based GIS and PGIS adhere to the same geospatial data accessibility, collaboration, and community participation concepts. Combining the two methodologies enables the development of interactive and participatory online mapping applications, allowing communities to actively participate in data collecting, analysis, and decision-making.

6.5.2   Mobile GIS

Mobile GIS is the use of mobile devices, such as smartphones and tablets, to collect, analyze, and visualize geospatial data in the field. Mobile GIS allows for real-time data collection, GPS tracking, and field mapping, making it useful for a wide range of applications such as environmental monitoring, asset management, and disaster response.

Relationship Between Mobile GIS and PGIS

- Community engagement: Mobile GIS apps can be created with user-friendly interfaces to engage people of all ages and backgrounds in data collection and mapping activities.
- Community empowerment: Mobile GIS tools can enable community members to actively participate in data collection, mapping, and decision-making. This is consistent with the key ideas of PGIS.
- Real-time data collection: Mobile GIS allows for real-time data collection in the field, which is useful for PGIS programs that require current information, such as disaster response or natural resource monitoring.
- Data accuracy: Mobile GIS can improve data accuracy by including GPS capabilities for precise georeferencing. This is especially significant in PGIS, where precise geographical data is critical to decision-making.
- Data sharing: Mobile GIS makes it easier to share field-collected data with other community members, researchers, and decision-makers, encouraging transparency and collaboration.
- Data visualization: Mobile GIS apps can show maps and spatial data on mobile devices, allowing community members to better view and analyze geospatial information, which is critical for participatory mapping projects.
- Local knowledge integration: PGIS frequently draws on local knowledge and community skills. Mobile GIS solutions enable community members to quickly contribute local knowledge by collecting observations and geographical data directly on their mobile devices.

In conclusion, mobile GIS and PGIS have a good relationship because they both prioritize community participation, local knowledge integration, and real-time data collecting. Integrating mobile GIS capabilities into PGIS programs can improve community empowerment, data accuracy, and the overall efficacy of participatory mapping and decision-making.

6.5.3 Crowdsourced Mapping

Crowdsourced mapping is a collaborative way to gathering geographical data from a diverse group of participants, frequently using the power of the crowd. Individuals or communities willingly share geographic information, such as GPS coordinates, images, and observations, which are then used to create maps and spatial databases.

Relationship Between Crowdsourced Mapping and PGIS

- Purpose: Crowdsourced mapping can serve a variety of purposes, including developing foundation maps for navigation and disaster response. PGIS is frequently used to meet specific community demands, such as land management and resource conservation.
- Scale: Crowdsourced mapping can function on a variety of scales, ranging from global initiatives such as OpenStreetMap to small-scale local operations. PGIS is typically community-focused and more localized in scope.
- Community involvement: Crowdsourced mapping and PGIS both place an emphasis on community involvement. Crowdsourced mapping allows individuals to contribute geospatial data, but PGIS actively involves communities in the mapping and data collection processes.
- Data quality: Due to the intimate engagement of communities, PGIS frequently prioritizes data quality and validation. Crowdsourced mapping platforms may use data quality control procedures on a larger scale.
- Data diversity: Crowdsourced mapping fosters different data contributions from a wide number of people, which enriches the knowledge accessible for decision making. PGIS stresses the value of local knowledge, ensuring that community perspectives are reflected in spatial data.
- Data ownership: PGIS and crowdsourced mapping may approach data ownership differently. PGIS frequently stresses community ownership and control over data, but crowdsourced mapping platforms may have different data ownership policies.
- Technology: Crowdsourced mapping frequently relies on mobile apps and web platforms to allow data entry from a large audience. PGIS may use similar technology, but it is usually community-centric and focused on local issues.
- Decision-making: Both approaches can help guide decision-making processes. PGIS takes into account local perspectives, whereas crowdsourced mapping might offer new data layers and context.

In brief, crowdsourced mapping and PGIS have similar principles of community interaction and data collection. While crowdsourced mapping spans a wide range of data sources and contributors, PGIS is a specialized technique that focuses on community-led geospatial efforts. The two share a dedication to using common knowledge and data for informed decision-making and community empowerment.

6.6 Future Directions of PGIS

GIS, especially participatory GIS, has exciting future possibilities and research opportunities as the field evolves. The following are some important areas of focus:

- *Community-centric GIS and PGIS*: The goal will be to ensure that GIS and PGIS efforts are really community-driven, with a focus on local knowledge, needs, and cultural sensitivity. The research will look into effective practices for community engagement and empowerment.
- *Indigenous knowledge and PGIS*: Research will continue to focus on the integration of indigenous knowledge and practices into PGIS efforts while upholding indigenous rights and sovereignty.
- *Ethical GIS and PGIS*: The research will look into the ethical implications of GIS and PGIS, including data privacy, informed consent, data ownership, and responsible data sharing. Developing ethical rules and procedures will be critical.
- *Equity and social justice*: A future study will look into how GIS and PGIS can be utilized to reduce social inequities and promote justice. This involves research on the geographical elements of social inequality as well as the role of geospatial technology in lobbying and policy reform.
- *Cross-sector collaboration*: GIS and PGIS will increasingly interact with other sectors, including public health, ecology, urban planning, and humanitarian aid. The study will investigate how interdisciplinary collaboration might lead to novel solutions to complicated challenges.
- *Community resilience and disaster management*: GIS and PGIS will play an important role in improving community resilience to catastrophes, with research focusing on real-time monitoring, early warning systems, and disaster recovery activities.
- *Web-based and mobile PGIS*: Scholars will look into expanding web-based and mobile PGIS applications, making it easier for communities to access and contribute to geospatial data from their smartphones and tablets.
- *Machine learning and AI in PGIS*: Incorporating machine learning and artificial intelligence into PGIS projects will improve data analysis, pattern identification, and predictive modeling, hence assisting community decision-making.
- *Community-driven data standards*: To ensure data interoperability, data standards and protocols for PGIS will be developed while taking into account community viewpoints and needs.
- *Geo-crowdsourcing in PGIS*: Through combining crowdsourcing techniques with PGIS, a broader community can participate in data collection and validation.
- *Inclusive GIS education*: Priority will be given to developing inclusive GIS and PGIS education and training programs that address accessibility, cultural diversity, and gender inclusion.
- *Policy advocacy*: The research will look into how GIS and PGIS can be utilized to influence policy decisions at the local, national, and international levels, advocating for community interests and environmental protection.

6.7   Conclusion and Arguments

In conclusion, it may be stated that GIS, particularly participatory GIS, is set for a future characterized by ethical data practices, community empowerment, and inter-disciplinary collaboration. As technology progresses, GIS must evolve to address social justice, environmental sustainability, and the unique demands of many communities. The concerns are enormous, ranging from protecting data privacy to fostering inclusion. However, GIS has enormous potential to help create a more equal, informed, and resilient world. By embracing ethical principles, encouraging cross-disciplinary relationships, and emphasizing community-driven initiatives, GIS can genuinely become a tool for everybody, as well as providing solutions to difficult global concerns.

Further Reading

Aitken, S., & Michel, S. M. (1995). Who contrives the "real" in GIS? Geographic information, planning and critical theory. *Cartography and Geographic Information Systems, 22*(1), 17–29. https://doi.org/10.1559/152304095782540519

Brown, G., & Kyttä, M. (2014). Key issues and research priorities for public participation GIS (PPGIS): A synthesis based on empirical research. *Applied Geography, 46*, 126–136. https://doi.org/10.1016/j.apgeog.2013.11.004

Bugs, G., Granell, C., Fonts, O., Huerta, J., & Painho, M. (2010). An assessment of public participation GIS and Web 2.0 technologies in urban planning practice in Canela, Brazil. *Cities, 27*, 172–181. https://doi.org/10.1016/j.cities.2009.11.008

Carver, S. (2003). The future of participatory approaches using geographic information: developing a research agenda for the twenty-first century. *Journal of the Urban and Regional Information Systems Association, 15*, 61–71.

Elwood, S. (2006). Critical issues in participatory GIS: Deconstructions, reconstructions, and new research directions. *Transactions in GIS, 10*(5), 693–708. https://doi.org/10.1111/j.1467-9671.2006.01023.x

Gottwald, S., Laatikainen, T. E., & Kyttä, M. (2016). Exploring the usability of PPGIS among older adults: Challenges and opportunities. *International Journal of Geographical Information Science, 30*(12), 2321–2338. https://doi.org/10.1080/13658816.2016.1170837

Gregory, I. N., & Ell, P. S. (2007). *Historical GIS: Technologies, methodologies, and scholarship.* Cambridge University Press. https://doi.org/10.1017/CBO9780511493645. ISBN: 9780511493645.

Huang, B. (2017). *Comprehensive geographic information systems.* Elsevier. ISBN: 9780128046609.

Kyem, P. (2004). Of intractable conflicts and participatory GIS applications; The search for consensus amidst competing claims and institutional demands. *Annals of the Association of American Geographers, 94*(1), 37–57. https://doi.org/10.1111/j.1467-8306.2004.09401003.x

Kyem, P. A. K., & Saku, J. C. (2009). Web-based GIS and the future of participatory GIS applications within local and indigenous communities. *The Electronic Journal on Information Systems in Developing Countries, 38*(7), 1–6. https://doi.org/10.1002/j.1681-4835.2009.tb00270.x

Ndzabandzaba, C. (2020). Participatory Geographic Information System (PGIS): A discourse toward a solution to traditional GIS challenges. In S. Brunn & R. Kehrein (Eds.), *Handbook of the changing world language map.* Springer. https://doi.org/10.1007/978-3-030-02438-3_122

Rambaldi G, Chambers R., McCall, M, And Fox, J. (2006). Practical ethics for PGIS practitioners, facilitators, technology intermediaries and researchers. PLA, IIED, London, 54,106–113.

Torres, M. D. L., González, R. D. M., & Yago, F. M. (2017). WebGIS and geospatial technologies for landscape education on personalized learning contexts. *ISPRS International Journal of Geo-Information, 6*(11), 350. https://doi.org/10.3390/ijgi6110350

Part II
Voices Mapping of Coastal Communities: Field Narratives, Case Studies, and Best Practices of Participatory GIS

Chapter 7
Importance and Scope of Voices Mapping in Coastal Communities: The Case of Coastal World

Maps are not just images; they are the stories of a place, the echoes of its people, and the hopes for its future.

Rebecca Solnit

Abstract In this chapter, we undergo a captivating journey into coastal neighborhoods, where a complex tapestry of human experiences and environmental issues intersect. *Voice mapping* acts as a compass, guiding us through the conceptual foundations that drive our investigation into the significance and extent of voice mapping in coastal communities. We explore the complex connection between community voices and geographic environments at the glocal (global and local) level, as well as laying the groundwork for a place of immersive scientific evidence. Therefore, we highlighted the necessity and extent of voice mapping for sustainability, as well as a glocal-level depiction of coastal communities.

Keywords Geospatial thinking · Community voices · Participatory mapping · Coastal world · Indian coastal society · Scientific evidence · Scope of voices mapping

In this chapter, we undergo a captivating journey into coastal neighborhoods, where a complex tapestry of human experiences and environmental issues intersect. *Voice mapping* acts as a compass, guiding us through the conceptual foundations that drive our investigation into the significance and extent of voice mapping in coastal communities. We explore the complex connection between community voices and geographic environments at the glocal level, as well as laying the groundwork for a place of immersive scientific evidence.

Key Points of the Chapter
- The importance and scope of voice mapping for sustainability.
- A representation of the voices of coastal communities at the glocal (global and local) scale.

7.1   Mapping Voices in Participatory Approaches: Concept and Significance

Participatory techniques highlight the necessity of including a variety of stakeholders in decision-making processes. In this context, *mapping voices* refers to a methodological tool for visualizing and recording the contributions, viewpoints, and experiences of individuals or groups. This technique is an important tool for identifying power dynamics, promoting inclusivity, and ensuring equal representation. Giving voice to excluded groups and publicizing their issues strengthens the democratic foundation of participatory processes, promoting more informed, inclusive, and long-term decision-making. In that instance, the application of participatory GIS tools and technologies assists the community in mapping their opinions from the ground.

7.2   Participatory Mapping for Inclusive Decision-Making Processes: Concepts and Linkages

Participatory mapping is a valuable tool for making inclusive decisions. This approach entails including stakeholders in the collaborative construction of maps that reflect their knowledge, preferences, and concerns about a certain area or subject. These maps serve as visible representations of various viewpoints, ensuring that all perspectives are considered during decision-making processes. The relationship between participatory mapping and inclusivity is clear and significant. By integrating a diverse set of stakeholders in mapmaking, decision-makers obtain a more thorough grasp of complicated situations, resulting in better informed and equitable decisions. This relationship is the foundation of effective and inclusive governance and community participation.

7.3   Geospatial Citizenship and Mapping Voices: A Geospatial Thinking

Geospatial thinking, as it relates to geospatial citizenship and mapping voices, highlights the importance of data literacy, equity and inclusivity, spatial awareness, and active community engagement in creating our communities and decision-making processes. The subsequent parts are going to look at how geospatial thinking intersects with geospatial citizenship and voice mapping. There are examples, such as the following:

- *Data literacy*: Geospatial thinking additionally demands data literacy, specifically the ability to analyze and use geographic information. Citizens must be able to access, comprehend, and interpret geospatial data, which may include maps, satellite imagery, and geographical statistics. Individuals with this level of literacy are better able to make educated decisions and advocate for their communities.

- *Spatial awareness*: Geospatial thinking begins with establishing spatial awareness. It encourages people to grasp their surroundings, both physically and socially. In terms of geospatial citizenship, this means being aware of the geographical features, resources, and challenges of one's neighborhood or region.
- *Community empowerment*: Geospatial thinking promotes communal empowerment. Citizens can actively shape policies and efforts affecting their lives by understanding the spatial elements of issues such as environmental degradation, resource availability, and urban development. This is consistent with the concept of geospatial citizenship, in which individuals play an active part in governance through spatial awareness and data-driven advocacy.
- *Equity and inclusivity*: Geospatial thinking encourages equitable and inclusive decision-making. It emphasizes the necessity of promoting varied viewpoints and addressing geographical inequities. Mapping voices becomes a tool for identifying and correcting spatial injustices, guaranteeing that no community is left behind.
- *Mapping voices*: Mapping voices is an essential part of geospatial citizenship. Individuals and communities' voices are collected, represented, and amplified utilizing geospatial tools and approaches. This can be accomplished through participatory mapping efforts that chronicle local knowledge and concerns on a geographic scale. It ensures that underprivileged voices are not excluded from decision-making processes.
- *Policy advocacy*: Geospatial thinking gives citizens the ability to advocate for evidence-based policies. Voice mapping and the use of spatial data to support ideas increase the legitimacy and persuasiveness of advocacy activities. Decision-makers are more likely to accept well-documented geographical evidence.

Finally, it may be argued that geographic thinking is critical to expanding geospatial citizenship and mapping voices. It enables residents to explore and actively engage in their communities' complex spatial aspects, promoting inclusivity, equity, and informed decision-making. This strategy is critical for creating resilient and sustainable societies in which all citizens' views are heard and valued.

7.4 Importance and Scope of Voice Mapping for Sustainability

In the context of participatory GIS, voice mapping is critical to improving sustainability and engaging geographical residents in meaningful ways. This novel technique encourages individuals to contribute their perspectives and expertise to the geospatial domain, resulting in a more inclusive and dynamic understanding of spatial dynamics. The value of voice mapping stems from its ability to magnify the voices of various populations, ensuring that their unique perspectives affect sustainable practices and policies. Participatory GIS with voice mapping improves spatial data accuracy and context by combining geospatial citizens' life experiences and local knowledge. This inclusiveness leads to a more comprehensive awareness of environmental concerns, facilitating the development of focused and successful sustainability efforts.

The scope of voice mapping goes beyond typical GIS approaches, providing a platform for community-driven decision-making. Geospatial citizens take an active role in the mapping process, affecting how spatial data is collected, analyzed, and used. This collaborative method not only democratizes geospatial discourse, but it also fosters a sense of ownership and responsibility among community members. Individuals can use voice mapping to express their worries, goals, and recommendations for sustainability, ensuring that solutions are not only scientifically sound but also culturally and socially sensitive. Furthermore, voice mapping proves to be an effective strategy for closing interactions and understanding gaps between communities and politicians. The participative character of this strategy promotes transparency and mutual respect, allowing for more effective discussion and cooperation. Geospatial citizens who actively map their surroundings acquire a stronger connection to their environment and a greater understanding of the interplay between human activities and ecological systems. This increasing awareness, combined with the democratization of spatial data, results in a more informed and ecologically conscientious population.

In conclusion, involving voice mapping in participatory GIS is critical for sustainability efforts. It converts geospatial citizens into active participants in the spatial discourse, supplementing data with varied perspectives and ensuring that sustainability activities are informed and inclusive. This strategy not only broadens the scope of GIS, but it also enhances the link between communities, their ecosystems, and decision-makers working toward a more sustainable future.

7.5 Voice Mapping and the Coastal World: Scientific Evidence

Voice mapping in the structure of participatory GIS has great promise for understanding and tackling the complex concerns of the coastal environment. Coastal regions are dynamic ecosystems shaped by the interaction of natural forces and human activity; therefore, community engagement is critical for long-term management. When combined with voice mapping in coastal situations, participatory GIS becomes a revolutionary tool, emphasizing collaboration and inclusivity. Voice mapping is about encouraging coastal communities to actively participate in mapping and documenting their local landscapes. Through this approach, geospatial citizens provide their direct knowledge, experiences, and observations about coastal dynamics. This democratization of spatial data means that the voices of those with deep ties to the coastal realm are integrated into the fabric of decision-making processes. This collaborative approach improves resilience by empowering communities to plan and implement long-term coastal management strategies.

In this regard, we compile (Table 7.1) scientific information from coastal worlds and characterize the representation of people or community-based voice recordings, mapping, and collections for further research, management, and decision-making processes.

Table 7.1 Voice mapping and the coastal world: scientific evidence

Sr. no.	Study purpose	Stakeholders/community-based voice recordings/ mapping	Investigated area(s)	References
1	Coastal zone planning	Variations in participatory forms are linked to their (stakeholders) differential powers Ground-based data collection with the help of people or community interaction	Norwegian coastal, Norway	Buanes et al. (2005)
2	Management of coastal zone	Illustrate the conditions and degree of stakeholder participation	Byfjorden in Bergen (Norway), Amvrakikos Gulf (Greece), and Nordre Älv Estuary (Sweden)	Oen et al. (2016)
3	Collaborative natural resource management	Developed community-based maps that encourage collaborative natural resource management	Luchun County, Yunnan in Southwest China	McConchie and McKinnon (2002)
4	Management of natural resource	Participatory spatial planning for community-based natural resource management	Cameroon (Africa)	McCall and Minang (2005)
5	Natural resource management	Integration of spatial information, GIS, and public participation	Coastal region of northern Australia	De Freitas (2010)
6	Management of natural resource	Integration of indigenous knowledge system and GIS	Review work from various sites in the world	Khan (2014)
7	Assessment of coastal flooding risk	Integration of GIS-based mapping and the role of people's capacity including social vulnerability	Bandar Abbas city region (Hormozgan province, Iranian south coast)	Hadipour et al. (2020)
8	Mapping vulnerability and pathways of viability	Participation of small scale fisheries to understand the pathways of viability	Sundarbans (India and Bangladesh)	Pattanaik (2021)
9	Understanding climate change perception	Perspectives of community-based management contexts	Colombia and Mexico	Ambrosio-Albala and Mar Delgado-Serrano (2018)

(continued)

Table 7.1 (continued)

Sr. no.	Study purpose	Stakeholders/community-based voice recordings/ mapping	Investigated area(s)	References
10	Integrated coastal zone management	Application of bi-directional knowledge sharing approach, participatory resource management approach, social learning approach	Thua Thien Hue (Vietnam)	Abelshausen et al. (2015)
11	Participatory coastal spatial planning	Application of participatory toolkit and mapping techniques	Coastal districts of Vietnam	Nguyen et al. (2017)
12	Marine policy development	Public participation	Global scales	da Silva et al. (2021)
13	Coastal restoration planning	Engagements of local knowledge through participatory modelling and active community participation	Coastal Louisiana	Hemmerling et al. (2019)
14	Monitoring ecosystem function	Engaging coastal community members for nature-based solutions approaches	Breton Sound Estuary (Northern Gulf of Mexico, USA)	Baustian et al. (2020)
15	Coastal resources and marine conservation	Public participation geographic information system mapping survey	Oregon coast at Cape Falcon, Cascade Head, Otter Rock, Cape Perpetua, and Redfish Rocks	Scully-Engelmeyer et al. (2021)
16	Management of coral reef ecosystem	Stakeholder identification and engagement through PGIS, and integrates the socio-spatial data on human-environment interactions	Coastal area of Hawaii Island	Levine and Feinholz (2015)
17	Disaster risk management	Application of an in-depth interviews with community persons and key informants and the PGIS methods	Oro Ombo (Indonesia)	Irawan et al. (2023)
18	Coastal decision-making processes	Application of public participatory GIS	Scotland (United Kingdom)	Green (2010)
19	Coastal zone management	Integration of integrated GIS methodology and community visualization for participatory approach	North Norfolk coast (England)	Jude et al. (2006)
20	Coastal biosphere reserve designation	Application of participatory GIS tools and techniques	Turneffe atoll (Belize)	Bridge Egan (2008)

7.6 Importance and Scope of Voice Mapping in Coastal Bengal, India

The coastal region of Bengal (Fig. 7.1) in eastern India is renowned for its intricate interplay of diverse ecosystems, communities, and socioeconomic activities. A comprehensive understanding of the dynamics of this area is imperative, particularly in light of the challenges posed by climatic vulnerability, socio-ecological changes, and dynamic shifts in land use patterns. Previous research in this field has unveiled a spectrum of contemporary socio-ecological challenges associated with climate change (Table 7.2). Positioned in the advanced landscape of social sciences research, especially within the framework of applying participatory techniques and leveraging geospatial thinking, the integration of stakeholder voice mapping with participatory GIS emerges as a potent instrument for policy planning. This approach introduces an innovative method for gathering and analyzing local knowledge, perspectives, and experiences, thereby contributing to the adept management and mitigation of issues within the region.

Considering the challenges and potential strengths of Bengal's coastal, the following importance and scope have been identified for further exploration in terms of participatory approaches:

Importance
- *Environmental monitoring and conservation*: Coastal Bengal is vulnerable to climate change consequences such as sea-level rise, cyclones, and erosion. Voice mapping can help to capture indigenous knowledge about environmental changes, allowing for more effective monitoring and conservation initiatives. Local knowledge about altering coasts, biodiversity hotspots, and susceptibility zones can be extremely important for sustainable resource management.
- *Disaster preparedness and response*: Given the region's vulnerability to natural disasters such as floods and cyclones, reliable and timely information is critical for successful disaster preparation and response. Voice mapping enables communities to provide real-time data on vulnerable regions, evacuation routes, and emergency services, thereby increasing the region's overall resilience.
- *Community engagement and empowerment*: Participatory GIS incorporates local communities into the mapping process, ensuring that their perspectives are heard and valued. By actively engaging residents in mapping their own environments, the technique fosters a sense of empowerment and ownership of the data and its implications.

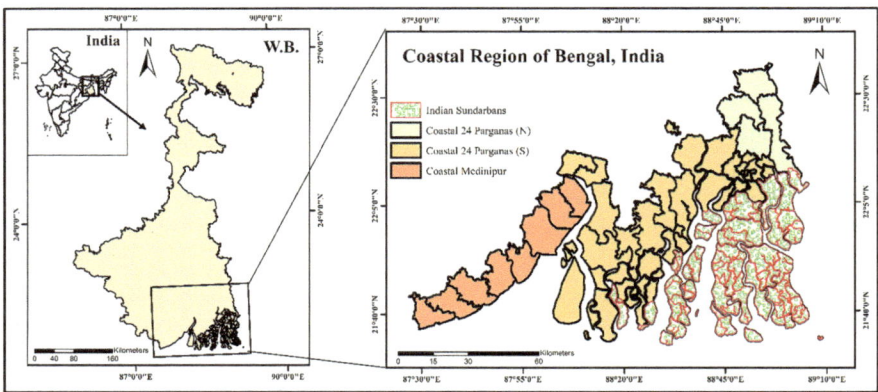

Fig. 7.1 The coastal region of Bengal, India. (Made by author(s))

Table 7.2 Climate change and socio-ecological challenges in coastal Bengal, India

Sr. no.	Study purpose	Contemporary socio-ecological challenges to climate change	Investigated area(s)	References
1	Vulnerability assessment	Assessment of long-term ecological vulnerability in terms of present and future climatic conditions	Indian Sundarban region	Sardar and Samadder (2023)
2	Socioeconomic vulnerability	Socioeconomic vulnerability to climate change	Indian Sundarbans	Sahana et al. (2021)
3	Environmental stress and livelihoods	Ecological stress and livelihood practices to climate change	Indian Sundarbans	Ghosh and Roy (2022)
4	Vulnerability assessment	Ecological and socioeconomic vulnerability to climate change	Gosaba Block, Indian Sundarbans	Mukherjee and Siddique (2021)
5	Mangroves' susceptibility to climate change	Sea level rise on mangrove dynamics	Indian Part of Sundarbans	Pramanik (2016)
6	Coastal land use/land cover	Multidimensional threats of degradation such as coastal land use/land cover	Indian Sunderbans	Datta and Deb (2012)
7	Coastal vulnerability assessment	Comparative exploration of coastal vulnerability	Purba Medinipur (West Bengal), Balasore (Odisha) coastal stretch	Hossain et al. (2022)
8	Coastal shoreline and land use/land cover dynamics	Transformation of land use/land cover aspects and shoreline changes	Coastal Medinipur (West Bengal)	Samanta and Paul (2016)
9	Assessment of societal vulnerability	Societal vulnerability of coastal regions	Sagar Island (West Bengal, India)	Mondal et al. (2020)
10	Empirical investigation of hazards and societal issues	Investigation on multi-hazard to climate change in coastal society	Indian Sundarban	Mondal et al. (2022)

- *Livelihood enhancement*: Coastal areas frequently rely on conventional occupations like fishing and agriculture. Understanding the spatial distribution of these activities with participatory GIS might help build targeted interventions to improve livelihoods. Furthermore, it can aid in the identification of alternate livelihood possibilities in response to shifting environmental conditions.

Scope
- *Interdisciplinary research*: The combination of participatory GIS and voice mapping allows for transdisciplinary study. Collaboration among environmental scientists, social scientists, and community members can provide comprehensive insights into the intricate interplay of natural systems and human activities in coastal Bengal.
- *Policy formulation*: Data obtained through participatory GIS can be used to develop evidence-based policies. Decision-makers can use voice mapping data to adjust policies to coastal communities' specific needs and concerns, supporting sustainable development.
- *Cultural heritage preservation*: Coastal Bengal is rich in cultural legacy, with numerous groups having their own rites and traditions. Integrating cultural mapping into participatory GIS can help preserve these intangible assets by instilling a sense of identity and continuity in local communities.

Voice mapping in coastal Bengal in India, through participatory GIS, is critical for tackling the region's diverse concerns. This strategy strengthens community resilience by amplifying local voices and integrating traditional knowledge with modern technologies, as well as contributing to sustainable development and informed decision-making.

7.7 Conclusion and Arguments

Voice mapping, especially when integrated with participatory GIS in coastal communities, highlights the significance of local knowledge in decision-making and sustainable development. Participatory GIS transforms into a powerful tool for empowering locals and ensuring that their unique viewpoints contribute to a more accurate picture of the coastal environment when they work together. This approach is essential in tackling climate change consequences, environmental deterioration, and socioeconomic vulnerabilities in coastal areas. Involving stakeholders in the mapping process allows for the identification of disaster-prone areas, monitoring of ecological changes, and the protection of cultural property. As well, the combination of participatory GIS and voice mapping promotes interdisciplinary research, encouraging collaboration among environmental scientists, social researchers, and local populations.

References

Abelshausen, B., Vanwing, T., & Jacquet, W. (2015). Participatory integrated coastal zone management in Vietnam: Theory versus practice case study: Thua Thien Hue province. *Journal of Marine and Island Cultures, 4*(1), 42–53. https://doi.org/10.1016/j.imic.2015.06.004

Ambrosio-Albala, D. P., & Mar Delgado-Serrano, D. M. (2018). Understanding climate change perception in community-based management contexts: Perspectives of two indigenous communities. *Weather, Climate, and Society, 10*(3), 471–485. https://doi.org/10.1175/WCAS-D-17-0049.1

Baustian, M. M., Jung, H., Bienn, H. C., Barra, M., Hemmerling, S. A., Wang, Y., et al. (2020). Engaging coastal community members about natural and nature-based solutions to assess their ecosystem function. *Ecological Engineering, 143*, 100015. https://doi.org/10.1016/j.ecoena.2019.100015

Bridge Egan, S. (2008). *Participatory GIS for biosphere reserve designation, Turneffe Atoll, Belize: Lessons learned.* Thesis submitted to the San Francisco State University. https://environment.sfsu.edu/sites/default/files/2022-05/Egan%252C%2520Stefanie_2008Thesis.pdf

Buanes, A., Jentoft, S., Maurstad, A., Søreng, S. U., & Karlsen, G. R. (2005). Stakeholder participation in Norwegian coastal zone planning. *Ocean & Coastal Management, 48*(9–10), 658–669. https://doi.org/10.1016/j.ocecoaman.2005.05.005

da Silva, M. Q., et al. (2021). *Public participation in marine policy development: Policy brief.* United Nations Educational, Scientific and Cultural Organization/UNESCO/IOC. https://unesdoc.unesco.org/ark:/48223/pf0000378082

Datta, D., & Deb, S. (2012). Analysis of coastal land use/land cover changes in the Indian Sunderbans using remotely sensed data. *Geo-spatial Information Science, 15*(4), 241–250. https://doi.org/10.1080/10095020.2012.714104

De Freitas, D. M. (2010). *The role of public participation, spatial information and GIS in natural resource management of the dry tropical coast, northern Australia.* Thesis submitted to the James Cook University, North Queensland, Australia. https://doi.org/10.25903/qh4c-yb42

Ghosh, S., & Roy, S. (2022). Climate change, ecological stress and livelihood choices in Indian Sundarban. In A. K. E. Haque, P. Mukhopadhyay, M. Nepal, & M. R. Shammin (Eds.), *Climate change and community resilience.* Springer. https://doi.org/10.1007/978-981-16-0680-9_26

Green, D. R. (2010). The role of Public Participatory Geographical Information Systems (PPGIS) in coastal decision-making processes: An example from Scotland, UK. *Ocean & Coastal Management, 53*(12), 816–821. https://doi.org/10.1016/j.ocecoaman.2010.10.021

Hadipour, V., Vafaie, F., & Deilami, K. (2020). Coastal flooding risk assessment using a GIS-based spatial multi-criteria decision analysis approach. *Water, 12*(9), 2379. https://doi.org/10.3390/w12092379

Hemmerling, S. A., Barra, M., Bienn, H. C., et al. (2019). Elevating local knowledge through participatory modeling: Active community engagement in restoration planning in coastal Louisiana. *Journal of Geographical Systems, 22*, 241–266. https://doi.org/10.1007/s10109-019-00313-2

Hossain, S. A., Mondal, I., Thakur, S., & Al-Quraishi, A. M. F. (2022). Coastal vulnerability assessment of India's Purba Medinipur-Balasore coastal stretch: A comparative study using empirical models. *International Journal of Disaster Risk Reduction, 77*, 103065. https://doi.org/10.1016/j.ijdrr.2022.103065

Irawan, L. Y., Devy, M. M. R., Prasetyo, W. E., Farihah, S. N., & Hartono, R. (2023). Taking the advantage of Participatory Geographic Information System (P-GIS) in validating Semeru disaster prone area map in Oro-Oro Ombo, Lumajang. In *IOP Conference Series: Earth and Environmental Science* (Vol. 1190, No. 1, p. 012006). IOP Publishing. https://doi.org/10.1088/1755-1315/1190/1/012006

Jude, S., Jones, A. P., Andrews, J. E., & Bateman, I. J. (2006). Visualisation for participatory coastal zone management: A case study of the Norfolk Coast. *England. Journal of Coastal Research, 22*(6), 1527–1538. https://doi.org/10.2112/04-0294.1

Khan, M. T. R. (2014). Geographical Information System (GIS) indigenous knowledge in natural resource management. *International Journal of Environment and Natural Sciences, 1*, 65–81.

Levine, A. S., & Feinholz, C. L. (2015). Participatory GIS to inform coral reef ecosystem management: Mapping human coastal and ocean uses in Hawaii. *Applied Geography, 59*, 60–69. https://doi.org/10.1016/j.apgeog.2014.12.004

McCall, M. K., & Minang, P. A. (2005). Assessing participatory GIS for community-based natural resource management: Claiming community forests in Cameroon. *Geographical Journal, 171*(4), 340–356.

McConchie, J. A., & McKinnon, J. M. (2002). Using GIS to produce community-based maps to promote collaborative natural resource management. *Asean Biodiversity, 2*, 27–34.

Mondal, M., Paul, S., Bhattacharya, S., & Biswas, A. (2020). Micro-level assessment of rural societal vulnerability of coastal regions: An insight into Sagar Island, West Bengal, India. *Asia-Pacific Journal of Rural Development, 30*(1–2), 55–88. https://doi.org/10.1177/1018529120946230

Mondal, M., Biswas, A., Haldar, S., Mandal, S., Mandal, P., Bhattacharya, S., & Paul, S. (2022). Climate change, multi-hazards and society: An empirical study on the coastal community of Indian Sundarban. *Natural Hazards Research, 2*(2), 84–96. https://doi.org/10.1016/j.nhres.2022.04.002

Mukherjee, N., & Siddique, G. (2021). Ecological and socio-economic vulnerability to climate change in some selected Mouzas of Gosaba block, the Sundarbans. In *Global Geographical Heritage, Geoparks and Geotourism: Geoconservation and Development* (pp. 105–129). https://doi.org/10.1007/978-981-15-4956-4_7

Nguyen, M. T., Tran, L. H., & Jhaveri, N. (2017). *Guide no. 2: Participatory mapping: Creating knowledge for coastal spatial planning in Vietnam. Toolkit for participatory coastal spatial planning at the district level.* USAID Tenure and Global Climate Change Program. https://www.land-links.org/wp-content/uploads/2018/03/USAID_Land_Tenure_TGCC_Vietnam_Toolkit_Guide_2.pdf

Oen, A. M., Bouma, G. M., Botelho, M., Pereira, P., Haeger-Eugensson, M., Conides, A., et al. (2016). Stakeholder involvement for management of the coastal zone. *Integrated Environmental Assessment and Management, 12*(4), 701–710. https://doi.org/10.1002/ieam.1783

Pattanaik, A. (2021). *Mangrove-dependent small-scale fisher (SSF) communities in the Sundarbans—Vulnerable yet viable.* Thesis submitted to University of Waterloo, Ontario, Canada. https://uwspace.uwaterloo.ca/bitstream/handle/10012/17366/Pattanaik_Aishwarya.pdf?sequence=3

Pramanik, M. K. (2016). Assessment of the impacts of sea level rise on mangrove dynamics in the Indian Part of Sundarbans using geospatial techniques. *Journal of Biodiversity, Bioprospecting and Development, 3*, 155. https://doi.org/10.4172/2376-0214.1000155

Sahana, M., Rehman, S., Paul, A. K., & Sajjad, H. (2021). Assessing socio-economic vulnerability to climate change-induced disasters: Evidence from Sundarban Biosphere Reserve, India. *Geology, Ecology, and Landscapes, 5*(1), 40–52. https://doi.org/10.1080/24749508.2019.1700670

Samanta, S., & Paul, S. K. (2016). Geospatial analysis of shoreline and land use/land cover changes through remote sensing and GIS techniques. *Modeling Earth Systems and Environment, 2*, 108. https://doi.org/10.1007/s40808-016-0180-0

Sardar, P., & Samadder, S. R. (2023). Long-term ecological vulnerability assessment of Indian Sundarban region under present and future climatic conditions under CMIP6 model. *Ecological Informatics, 76*, 102140. https://doi.org/10.1016/j.ecoinf.2023.102140

Scully-Engelmeyer, K. M., Granek, E. F., Nielsen-Pincus, M., & Brown, G. (2021). Participatory GIS mapping highlights indirect use and existence values of coastal resources and marine conservation areas. *Ecosystem Services, 50*, 101301. https://doi.org/10.1016/j.ecoser.2021.101301

Further Reading

Craig, W. J., Harris, T. M., & Weiner, D. (Eds.). (2002). *Community participation and geographical information systems*. CRC Press. ISBN 9780367578626.

Elwood, S. (2002). GIS use in community planning: A multidimensional analysis of empowerment. *Environment and Planning A, 34*, 905–922. https://doi.org/10.1068/a34117

Lewis, D. M. (1995). Importance of GIS to community-based management of wildlife: Lessons from Zambia. *Ecological Applications, 5*(4), 861–871. https://doi.org/10.2307/2269337

Solís, P., & Zeballos, M. (Eds.). (2022). *Open mapping towards sustainable development goals: Voices of YouthMappers on community engaged scholarship*. Springer. https://doi.org/10.1007/978-3-031-05182-1

Tsai, B. W., Chang, C. Y., Lin, C. C., & Lo, Y. C. (2002). Public participation geographic information system and indigenous society: New partnership of indigenous peoples in Taiwan. *Geography Research Forum, 26*, 152–163 https://grf.bgu.ac.il/index.php/GRF/article/view/311

Chapter 8
Voice Mapping of Coastal Communities: Field Narratives from the Coastal Medinipur and Sundarbans

In every outthrust headland, in every curving beach, in every grain of sand, there is the story of the earth.

Rachel Carson

Abstract Within the expansive canvas of our planet, coastal communities stand as singular witnesses to the dynamic interaction between humanity and the sea. This chapter invites readers on a journey into the essence of coastal environments, capturing the voices that echo in the rhythm of daily life. The coastal regions of Medinipur and the Sundarbans were examined, where the waves carry with them the accumulated wisdom of generations. In this regard, these field narratives serve as beacons, highlighting intricate stories of resilience, adaptation, and symbiosis on the coasts. Also discussed were human-mangrove conflicts, man-environment linkages, socio-ecological transition, challenges, conservation, sustainable marine fishing techniques, and the extent of climate gap management among climate migrants. This compilation is more than just documentation; it is an ode to the voices that are frequently forgotten and drowned in the thunder of the seas. Through the prism of the field accounts offered here, we urge readers to join us on a profound exploration of the social, cultural, ecological, and human components that compose the coastal substance.

Keywords Community voices · Socio-ecological relationship · Field narratives · Management of coastal environments · Thinking of participatory GIS · Coastal region of Bengal

Within the expansive canvas of our planet, coastal communities stand as singular witnesses to the dynamic interaction between humanity and the sea. This chapter invites readers on a journey into the essence of coastal environments, capturing the voices that echo in the rhythm of daily life. The coastal regions of Medinipur and the Sundarbans (Fig. 8.1) were examined, where the waves carry with them the accumulated wisdom of generations. In this regard, these field narratives serve as beacons, highlighting intricate stories of resilience, adaptation, and symbiosis on

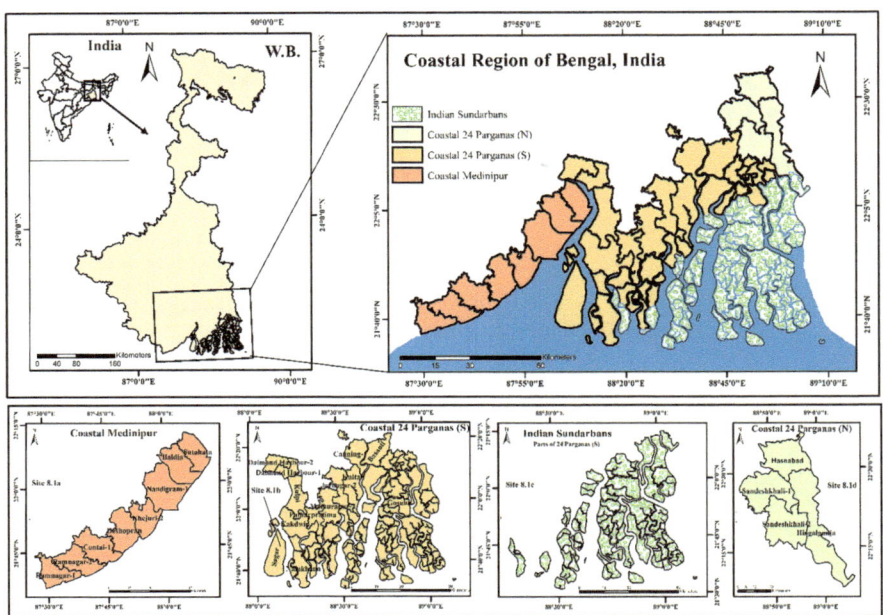

Fig. 8.1 Locational space of the study area (coastal region of Bengal, India). Site 8.1(**a**) Coastal region of Medinipur. Site 8.1(**b**) Sagar Island (Coastal South 24 Parganas). Site 8.1(**c**) Indian Sundarbans. Site 8.1(**d**) Coastal North 24 Parganas (all are part of the coastal region of Bengal, India). (Made by authors)

the coasts. Also discussed were human-mangrove conflicts, man-environment linkages, socio-ecological transition, challenges, conservation, sustainable marine fishing techniques, and the extent of climate gap management among climate migrants. Furthermore, life's rhythm in these seaside enclaves is a timeless song that combines the trials and accomplishments of the communities that live there. The stories told in these settings range from mundane to remarkable, revealing the complex dance between humans and nature. This compilation is more than just documentation; it's a tribute to often-overlooked voices lost in the sea's thunder. Through these field stories, we ask readers to embark on a thorough examination of the social, cultural, ecological, and human factors that influence coastal life. In embarking on this journey, we recognize the significant contributions of the individuals, families, and communities who graciously shared their tales. Their voices reverberate across these pages, imploring us to pay attention and comprehend the fragility and resilience embedded in the fabric of coastal life.

Key Points of the Chapter
- Exploring the breadth of voices documented in field narratives within coastal communities.
- Explains how humans and nature are related in a socio-ecological way.
- Field narratives on the following:

- Human-mangrove conflicts in the Sundarbans.
- Man-environment relationships in the coastal region of Medinipur.
- Socio-ecological transformation, challenges, and conservation of the Sundarbans.
- Sustainable thinking in the traditional marine fishing community on the Medinipur Coast.
- Scope of climate gap management among climate migrants on Sagar Island.

8.1   Introduction

The data collection (Fig. 8.2) and mental mapping components of participatory GIS approaches include voice mapping of coastal communities via field narratives. We have the option to look at it as first-hand narratives, anecdotes, or descriptions of experiences and observations obtained during fieldwork. The ensuing discussions form a collection of field narratives originating from the coastal regions of Medinipur and the Sundarbans. These narratives encompass topics such as conflicts between humans and mangroves, relationships between humans and the environment, socio-ecological transformation, challenges faced, and sustainable conservation thinking within the traditional marine fishing community. Additionally, the study explores the scope of climate gap management among climate migrants. The following three stages were followed for in-depth explorations of the narratives, such as the following:

Stage 1: Field narratives from the ground
Stage 2: A process to integrate with scientific evidence
Stage 3: The moral of this journey

8.2   Field Interactions Unveiled: A Compilation of Field Narratives

8.2.1   Coastal Dialogues: Investigating Human-Mangrove Conflicts Through the Participatory GIS

Field Narrative
The Sundarbans (Fig. 8.1, Site 8.1c), a UNESCO World Heritage Site with blooming mangrove forests, are also a place where the delicate balance between environment and human societies is continuously strained (Bandyopadhyay, 2016; UNESCO, 1997). In our field narrative, we use participatory methodologies (Table 8.1) to investigate the complicated subject of mangrove-human conflicts among Sundarbans' traditional marine fishermen. The coastal dialogues emerged as an intrigue and connection story in the heart of India's Sundarbans, where

Fig. 8.2 Gathering data and field narratives from the coastal areas of Medinipur and the Sundarbans in India

Table 8.1 Participatory methodologies: functional details of field observations

Location	Time duration	Participants[a]
Coastal Namkhana, Pathar Pratima, Kultali, Gosaba, and Hingalganj (West Bengal, India) *Used approaches/techniques*: FGDs, HH conversation, and thematic analysis	April to June 2022	1. Mr. C. Sarkar (Age: 55+) 2. Md. Ajijul (Age: 47+) 3. Sk. Masud (Age: 60+) 4. Mr. Raju (Age: 55+) 5. Mrs. Pravati (Age: 36+) 6. Ms. Krishna (Age: 24+)

Note: Field-visiting data by the author (K.D. Malakar)
[a]Name changed

mangroves and human settlements cohabit in a delicate tango. As I navigated the convoluted waterways, I found a world in which the whispers of the tides held stories of resilience and war. I met with local residents and heard stories about shifting livelihoods and vanishing mangrove cover. A seasoned fisherman detailed the declining fish stocks, and connecting his story with the ebb and flow of the tide. I also observed several direct and indirect actions relating to the relationship between humans and mangroves, as well as the disputes that arise. As a result, the following conflicts and interventions have been included in the story:

Livelihood Purpose
Conflict/interventions: Livelihoods, especially forestry and fishing activities, influenced the mangrove environment.
Perception/narratives: "During our fishing trips, particularly when picking honey from the forest, several inappropriate acts, such as chopping young seedlings, toppling enormous tree stalks, and cutting mangroves, may result in intervention." (Original in Bengali: "Āmādēra mācha dharāra bhramaṇēra samaẏa, biśēṣa karē bana thēkē madhu tōlāra samaẏa, bēśa kichu anupayukta kāja, yēmana taruṇa cārā kāṭā, biśāla gāchēra ḍālapālā upaṛē phēlā ēbaṁ myānagrōbha kāṭāra phalē hastakṣēpa hatē pārē.")

Sources of Traditional Medicine
Conflict/interventions: The majority of locals believe in and practice traditional mangrove medicine. However, the gathering procedure (tree bark and leaves) is unscientific.
Perception/narratives: "While gathering mangrove tree bark and leaves for medicinal purposes, we often cut down and collect the small seedlings." (Original in Bengali: "Auṣadhi uddēśyē myānagrōbha gāchēra chāla ēbaṁ pātā saṅgraha karāra samaẏa, āmarā prāẏaśa'i chōṭa cārā kēṭē saṅgraha kori.")

Housing Materials
Conflict/interventions: Poor households in the village utilize mangrove logs to build their houses.
Perception/narratives: "In our village, most homes use mangrove wood, though its use in government-provided brick houses has slightly declined. However, it's still significantly less than before, now allocated for diverse human needs like

channel bulwarks. We collect both immature and mature wood from convenient locations." (Original in Bengali: "Āmādēra grāmē, bēśirabhāga bāṛitē'i myānagrōbha kāṭha byabahāra karā haẏa, yadi'ō sarakāra-pradatta iṭēra bāṛitē ēra byabahāra kichuṭā kamēchē. Yā'ihōka, ēṭi ēkhana'ō āgēra tulanāẏa ullēkhayōgyabhābē kama, ēkhana cyānēla bālōẏārkēra matō bibhinna mānabika praẏōjanēra jan'ya barādda karā haẏēchē. Āmarā subidhājanaka abasthāna thēkē aparipakka ēbaṁ paripakka ubhaẏa kāṭha saṅgraha kori.")

Shrimp Farming
Conflict/interventions: Mangroves are being cleared from the bank, and shrimp farming is being established as a source of revenue.
Perception/narratives: "We built and are developing shrimp farming in the coastal area as a source of revenue. However, the expansion of shrimp aquaculture has contributed to the destruction of coastal mangroves." (Original in Bengali: "Āmarā upakūlīẏa ēlākāẏa ciṇṛi cāṣa tairi karēchi ēbaṁ rājasbēra uṭsa hisēbē unnaẏana karachi. Yā'ihōka, ciṇṛi jalaja cāṣēra samprasāraṇa upakūlīẏa myānagrōbha dhbansē abadāna rēkhēchē.")

In addition, the growth of agriculture, settlement expansion, unplanned tourism, and anthropogenic-induced climate vulnerabilities was identified, all of which had an impact on mangrove natural ecosystems. As a result, the notion of human-mangrove conflict is addressed here through both local perception and field observations.

A Process to Integrate with Scientific Evidence
Based on the collected narratives, the researcher may integrate it with the science by following the possible stages, such as the following:

Initially, comprehensive field surveys gather qualitative narratives from local communities, documenting their experiences and insights. At the same time, scientific information about climatic variability, sea-level rise, human intervention, and mangrove health can be gathered. And during the integration phase, participatory GIS techniques may be used to overlay community narratives on scientific maps. The combination of qualitative narratives and quantitative evidence ensures a thorough understanding of the complex dynamics at work, establishing the groundwork for informed conservation policies and sustainable coexistence in the vulnerable coastal area (Bwambale, 2023; Zidny et al., 2020).
Furthermore, this synthesis identifies conflict hotspots by linking anecdotal evidence to quantitative ecological shifts. The combination of local knowledge and scientific evidence not only broadens our understanding of the complex relationships but also serves as the foundation for informed conservation plans adapted to the Sundarbans' particular dynamics.

The Moral of This Journey
The Sundarbans Coastal Dialogues emphasize the importance of combining science and indigenous wisdom to find successful answers. It serves as a reminder that sustainable coexistence necessitates a collaborative harmony in which both nature and the community contribute to the conservation symphony.

8.2.2   Man-Environment Relationships: Shifting Shores, Shifting Stories

Field Narrative

Shifting Shores, Shifting Stories emerged as an engrossing voyage into the complex dance of man-environment relationships along the coastal stretches of Medinipur (Fig. 8.1, Site 8.1**a**), where the confluence of land and sea portrayed a picture of perpetual change. As a researcher, I entered the maze of fishing communities and found a mosaic of tales carved into the sands of time. As well, Table 8.2 highlights the functional details of field observations. The perceptions and narratives in this regard are listed below:

- Nestled among coconut palms in the heart of a village, a group of women gathered to share poignant tales. Their narratives painted vivid pictures of ancestral homes succumbing to the relentless advance of the waves, skillfully weaving memories into the ever-changing fabric of the coastline.
- An experienced cultivator saw connections between the ever-changing climate and the volatility he witnessed during his seasonal harvests.
- A seasoned fisherman described the shifting fish populations, attributing them to elusive ocean currents, sea-level rise, and varying water temperatures.
- One of the respondents stated that the increased cyclonic activity on the Medinipur coast, as well as the associated risks, had an impact on the fishermen's everyday lives and livelihoods, including their eating preferences and other sociocultural activities.

Therefore, the narratives were transformed into bright visual stories that reflected the ebb and flow of life along the coastlines via participatory communication activities. *Shifting Shores, Shifting Stories* emerged not just as a record of environmental changes but also as a live history of the indomitable spirit of coastal communities in Medinipur as they tried to adapt to the whims of their ever-changing surroundings. The initiative served as a bridge between the tangible facts of environmental transformations and the intangible yet poignant stories carved in the psyches of coastal people.

Table 8.2 Participatory methodologies: functional details of field observations

Location	Time duration	Participants
Coastal Digha, Shankarpur, Tajpur, Silampur, and Mandarmani (West Bengal, India) *Used approaches/techniques*: FGDs, HH conversation, and thematic analysis	August to September 2023	Total respondents: 33 Not willing to provide the respondent's name Age groups: Below 18 (Yrs.): 7 Above 18 (Yrs.): 26

Note: Field-visiting data by the author (S. Roy)

A Process to Integrate with Scientific Evidence
Field surveys collect rich narratives from coastal communities, combining personal tales with environmental changes. Coastal researchers may utilize scientific data on sea level rise, temperature variations, and socio-ecological changes for further investigation in conjunction with local perspectives and technical knowledge in policy planning (Yu & Mu, 2023; Avilés Irahola et al., 2022; Hermans et al., 2022; Kettle et al., 2014). During the integration phase, interactive workshops and mapping sessions will bring these parts together, resulting in a visual tapestry in which local stories coincide with empirical information.

The Moral of This Journey
The journey tells us that true wisdom comes at the intersection of scientific investigation and time-honored storytelling when we understand the growing link between humans and their coastal environment.

8.2.3   Coastal Diaries: Capturing the Journey of Transformation, Challenges, and Conservation in the Sundarbans

Field Narrative
In the heart of the Sundarbans (Fig. 8.1, Site 8.1c), where land meets the sea, we embarked on a journey that would become a story of socio-ecological transformation, obstacles, and conservation. Armed with cameras and notebooks, we set out to chronicle the coastal treasures that distinguish the Sundarbans as a distinct ecosystem. Challenges arose when the researchers met the constant dangers to this delicate shelter. Rising sea levels and human encroachment pose serious threats to the Sundarbans. With each passing day, the researcher crossed tight canals, unearthing stories of populations living on the brink, confronted with both nature's unpredictability and human damage. Table 8.3 discusses the functional details of field observations in this regard. The researchers met devoted individuals and groups working tirelessly to safeguard this UNESCO World Heritage Site, and they noticed the following substances:

Table 8.3 Participatory methodologies: functional details of field observations

Location	Time duration	Participants
Coastal 24 Parganas (West Bengal, India) *Used approaches/techniques*: FGDs, HH conversation, and thematic analysis	August to September 2023	Total respondents: 12 Not willing to provide the respondent's name Age groups: Below 18 (Yrs.): 2 Above 18 (Yrs.): 10

Note: Field-visiting data by the author (S. Roy)

- One respondent indicated that rising sea levels had harmed coastal fishing grounds in mangrove regions as well as agricultural land areas during high tides and cyclonic activity.
- One community, identified as climate migrants, relocated from their ancestral territory to Sagar Island to protect their lives and fundamental livelihoods.
- Other socio-ecological issues identified during the field investigations include a decline in fishing species, coastal erosion, shoreline alteration, and socio-ecological fragility, all of which contribute to socio-ecological transformation.
- On a lighter note, we examine community ways to conserve in a sustainable manner, "bottom-up spatial approaches." Only one recommendation has been made in this regard.

A Process to Integrate with Scientific Evidence
The incorporation of community involvement into the preservation and sustainability of the coastal socio-ecological system in the Sundarbans is one of the innovative initiatives for restoring spatial sustainability in the face of the climate disaster. In this regard, we propose the following ways for development and sustainability:

(a) Mangrove conservation (Tomaquin, 2023; Damastuti & de Groot, 2019; Astuti et al., 2018).
(b) Climate-resilient agriculture (Das et al., 2022; Manda et al., 2019; Sanogo et al., 2017).
(c) Saltwater intrusion monitoring (Van Tuan et al., 2023; and Kallioras et al., 2006).
(d) Biodiversity hotspot conservation (Scully-Engelmeyer et al., 2021; Ernoul et al., 2018).
(e) Integrated disaster preparedness (Liu et al., 2018; and Reichel & Frömming, 2014).
(f) Community-driven water management (de Carvalho et al., 2021; Ismail et al., 2019; Panek & Van Heerden, 2013; and Aabeyir et al., 2012).
(g) Community-based climate education and justice (Haklay & Francis, 2018).
(h) Eco-tourism for sustainability (Yasin & Woldemariam, 2023; Rahman, 2010).
(i) Participatory mapping workshops (Bakhshianlamouki et al., 2023; Tourinho et al., 2023; Mere-Roncal et al., 2021).
(j) Policy advocacy for community involvement (Rawat & Yusuf, 2020; Duval-Diop et al., 2010; Di Gessa, 2008).
(k) Thinking of sustainability and practices (de Carvalho et al., 2021; Hedelin et al., 2017).

The Moral of This Journey
The Sundarbans field journey reminds us that togetherness breeds resilience. In the face of change and challenges, conservation becomes more than a duty but a community obligation, echoing the critical truth that our individual and collective actions impact the fate of our vulnerable coastal ecosystems.

8.2.4   Whispers of the Estuary: Sustainable Thinking in the Traditional Marine Fishing Community

Field Narrative

A remarkable story evolved in the coastal villages of Medinipur (Fig. 8.1, Sites 8.1**a** and Table 8.4), where the estuary murmured tales of the sea. It was sustainable thinking in a traditional marine fishing community. As dawn bathed the sky in orange, fishermen set sail in colorful boats, gliding through the complicated network of estuary waterways. The narrative began with the rhythmic draw of fishing nets, a time-honored dance between man and nature. However, a small movement was noticeable. This traditional village has embraced a new ethos: sustainable fishing. They recognized the delicate balance required to protect marine biodiversity while maintaining their livelihoods. As the day continued, the whispers of the estuary revealed the community's troubles. Overfishing and climate change threaten their way of life. The fishermen, who had previously relied exclusively on the day's catch, were now forced to adopt more sustainable tactics. They learned to navigate the fluctuating tides using traditional knowledge and contemporary conservation measures. The community's innovative ideas were central to the narrative. Locally made fish-aggregating devices, selective fish catching, management of coastal shores, monitoring and cooperation with the nearest fishing department, and selective fishing gear reduce bycatch, promoting a balance between human needs and environmental well-being. The estuary became a classroom, and the fishermen became stewards, realizing that sustainable thinking was not an option but a requirement. In the shadow of tree-planting-fringed coasts, the community formed a marine conservation committee to symbolize a sense of shared responsibility. They educated future generations, passing on not only fishing techniques but also a heritage of sustainable coexistence with the estuary.

The story ended with a scene of unity as the community came together to enjoy their sustainable crop. The estuary whispered its approval, a tribute to the power of combining tradition and innovation, demonstrating that even in the immensity of the sea, little whispers of change may resound deeply, determining the future of coastal communities.

Table 8.4 Participatory methodologies: functional details of field observations

Location	Time duration	Participants
Coastal Medinipur (West Bengal, India) *Used approaches/techniques*: FGDs, HH conversation, and thematic analysis	August to September 2023	Total respondents: 10 Not willing to provide the respondent's name Age groups: Below 18 (Yrs.): 0 Above 18 (Yrs.): 10

Note: Field-visiting data by the author (S. Roy)

A Process to Integrate with Scientific Evidence
In collaboration with marine biologists and ecologists, we may explore the local and community knowledge integration of sustainable fishing practices. Interviews with community members provided qualitative insights, complementing quantitative data. Continuous dialogue with experts may ensure alignment with scientific principles. These collaborations assist in promoting new and innovative sustainable spatial practices for management and sustainable development (Piñeiro-Corbeira et al., 2022; Berkström et al., 2019; De Freitas & Tagliani, 2009; Valbo-Jørgensen & Poulsen, 2000).

The Moral of This Journey
Sustainability (especially the "thinking of sustainability") is the hyperlink between tradition and a prosperous future. Communities can ensure that the delicate rhythms of the estuary are preserved for future generations by combining ancient wisdom with modern science.

8.2.5 Scope of Climate Gap Management Among Climate Migrants on Sagar Island (India)

Field Narrative
On Sagar Island (Fig. 8.1, Sites 8.1**b**), where the Ganges (Hooghly River) meets the Bay of Bengal, a story of resilience and adaptation unfolds within the climate divide (Table 8.5). The island, which is vulnerable to rising sea levels and violent weather, experienced a unique experiment in climate gap management among its migrants. The story began with the family being forced to abandon their ancestral homes due to rising water levels. Nevertheless, a sense of community grew despite the dislocation. Locals, using sustainable building techniques, welcomed the island's shifting topography, as well as marking the beginning of a new chapter in the story. The characters were climate migrants who combined old knowledge with modern technology to create a tapestry of survival. The story revealed innovative housing solutions that used eco-friendly materials and resilient architecture to protect against the harsh effects of the climate. The community-led activities were central to the narrative. Collaborations with NGOs and government agencies facilitated instructional initiatives on innovative fishing practices, community-based water management, climate-resilient agriculture, and alternative livelihoods. The islanders went from victims to agents of change, mastering climate gap management.

Table 8.5 Participatory methodologies: functional details of field observations

Location	Time duration	Participants
Sagar Island (West Bengal, India) *Used approaches/techniques*: FGDs, HH conversation, and thematic analysis	August to September 2023	Total respondents: 8 Not willing to provide the respondent's name Age groups: Below 18 (Yrs.): 2 Above 18 (Yrs.): 6

Note: Field-visiting data by the author (S. Roy)

As the story continued, the migrants evolved into storytellers, sharing their adventures with the globe. The islanders on Sagar Island transformed into a beacon of hope in this socio-ecologically transformative society, demonstrating that effective climate gap management needs both adaptation and empowerment. The island, which was previously devastated by climate change, has become a practice model for sustainable living, but many contemporary issues are still there, like cyclonic hazards (Paul et al., 2024; Bhattacharjee & Dhara, 2021), coastal floods (Das, 2023; Gopinath, 2010), sea-level rise during high tides (Chakraborty & Saha, 2020), coastal erosion (Gopinath & Seralathan, 2005), various coastal vulnerabilities (Mondal et al., 2020), and related others. The epilogue suggested a future in which climate migrants, equipped with knowledge and solidarity, could bridge the gaps left by a changing environment, such as local ecological knowledge and thinking of sustainability.

Sagar Island's story was a tribute to humanity's indomitable spirit, demonstrating that within the folds of climatic gaps, there is room for ingenuity, adaptation, and a common commitment to forging a sustainable future.

A Process to Integrate with Scientific Evidence
The incorporation of scientific knowledge into the scope of climate gap management required a methodical approach. Initial engagement with climatologists and environmental scientists may have facilitated in-depth research and a more scientific strategy for future development in the context of the climate issue. In this regard, on-site scientific quantitative data collection and research using participatory GIS tools will aid in policy planning for the reduction of climate-induced vulnerabilities. As a result, purposeful integration is recommended, and academics are invited to conduct additional research and policy planning for sustainable climate gap management. Moreover, this technique is intended to increase the story's believability by anchoring it in scientific reality. By incorporating scientific evidence into the narrative, Sagar Island's story evolved from a personal experience to a data-driven investigation of climate gap management in the face of environmental threats.

The Moral of This Journey
The expedition to Sagar Island underscores the ability of communities, equipped with knowledge and teamwork, to overcome challenges and unite in bridging the climate divide. The islanders illustrated that taking proactive measures for climate management is a collective responsibility toward ensuring a sustainable future.

8.3 Conclusion and Arguments

In conclusion, the stories from Coastal Medinipur and the Sundarbans depict a complex tapestry of problems, resilience, and community-driven solutions. These coastal communities' views highlight the critical need for sustainable practices in the face of environmental concerns. The arguments offered in these field narratives emphasize the connection between human experiences and environmental concerns.

They emphasize the need of preserving traditional knowledge, encouraging community-led conservation, and building resilience to climate change. These stories promote a comprehensive approach in which local voices, scientific evidence, and collaborative efforts come together to determine a sustainable future for these coastal environments.

References

Aabeyir, R., Kabo-bah, A. T., & Campus, P. O. (2012). The Role of public participatory gis in rural water resources mapping. In *Proceeding of The IASTED 2012 African Conferences* (pp. 3–5). Gaborone. https://doi.org/10.2316/P.2012.762-027

Astuti, S., Muryani, C., & Rindarjono, M. G. (2018). The community participation on mangrove conservation in Sayung, Demak from 2004–2016. In *IOP Conference Series: Earth and Environmental Science* (Vol. 145, No. 1, p. 012087). IOP Publishing. https://doi.org/10.1088/1755-1315/145/1/012087

Avilés Irahola, D., Mora-Motta, A., Barbosa Pereira, A., et al. (2022). Integrating scientific and local knowledge to address environmental conflicts: the role of academia. *Human Ecology, 50,* 911–923. https://doi.org/10.1007/s10745-022-00344-2

Bakhshianlamouki, E., Augustijn, E. W., Brugnach, M., Voinov, A., & Wijnberg, K. (2023). A participatory modelling approach to cognitive mapping of the socio-environmental system of sandy anthropogenic shores in The Netherlands. *Ocean & Coastal Management, 243,* 106739. https://doi.org/10.1016/j.ocecoaman.2023.106739

Bandyopadhyay, K. (2016). Civilisation in Sunderbans traced to Mauryan era. *Times of India* (Kolkata), Date 1 August 2016 and Retrieved 8 December 2023. Available at: https://timesofindia.indiatimes.com/city/kolkata/civilisation-in-sunderbans-traced-to-mauryan-era/articleshow/53483794.cms

Berkström, C., Papadopoulos, M., Jiddawi, N. S., & Nordlund, L. M. (2019). Fishers' local ecological knowledge (LEK) on connectivity and seascape management. *Frontiers in Marine Science, 6,* 130. https://doi.org/10.3389/fmars.2019.00130

Bhattacharjee, A., & Dhara, S. (2021). Natural hazards and disasters: Management at Sagar Island of South 24 Parganas district. *International Journal of Creative Research Thoughts, 9*(1), 4571–4578. https://ijcrt.org/papers/IJCRT2101557.pdf

Bwambale, B. (2023). Integrating indigenous knowledge with science to suitably tackle disasters due to climate and environmental change: An overview of the progress and way forward. In *Multi-hazard vulnerability and resilience building* (pp. 127–143). https://doi.org/10.1016/B978-0-323-95682-6.00008-5

Chakraborty, D., & Saha, P. (2020). Vulnerability assessment for Sagar Island coast. *West Bengal with Respect to Inundation Hazards, 49*(9), 1521–1527. https://nopr.niscpr.res.in/bitstream/123456789/55525/1/IJMS%2049(9)%201521-1527.pdf

Damastuti, E., & de Groot, R. (2019). Participatory ecosystem service mapping to enhance community-based mangrove rehabilitation and management in Demak, Indonesia. *Regional Environmental Change, 19,* 65–78. https://doi.org/10.1007/s10113-018-1378-7

Das, K. K. (2023). Flood hazard mapping of Sagar Island during cyclone 'YAAS' using remote sensing and GIS. *Journal of Geography & Natural Disasters, 13*(2), 272.

Das, U., Ansari, M. A., & Ghosh, S. (2022). Effectiveness and upscaling potential of climate smart agriculture interventions: Farmers' participatory prioritization and livelihood indicators as its determinants. *Agricultural Systems, 203,* 103515. https://doi.org/10.1016/j.agsy.2022.103515

de Carvalho, C. M., Luiz Giatti, L., Fagerholm, N., Bedran-Martins, A. M., & Kytta, M. (2021). Participatory Geographic Information Systems (PGIS) to assess water, energy and food availability in a vulnerable community in Guarulhos (Brazil). *International Journal of Urban Sustainable Development, 13*(3), 516–529. https://doi.org/10.1080/19463138.2021.2019041

De Freitas, D. M., & Tagliani, P. R. A. (2009). The use of GIS for the integration of traditional and scientific knowledge in supporting artisanal fisheries management in southern Brazil. *Journal of Environmental Management, 90*(6), 2071–2080. https://doi.org/10.1016/j.jenvman.2007.08.026

Di Gessa, S. (2008). Participatory Mapping as a tool for empowerment: Experiences and lessons learned from the ILC network. *Knowledge for Change' Series (International Land Coalition)*, 1–53. https://dlc.dlib.indiana.edu/dlc/bitstream/handle/10535/3647/08_ILC_Participatory_Mapping_Low.pdf?sequence=1&isAllowed=y

Duval-Diop, D., Curtis, A., & Clark, A. (2010). Enhancing equity with public participatory GIS in hurricane rebuilding: Faith based organizations, community mapping, and policy advocacy. *Community Development, 41*(1), 32–49. https://doi.org/10.1080/15575330903288854

Ernoul, L., Wardell-Johnson, A., Willm, L., Béchet, A., Boutron, O., Mathevet, R., et al. (2018). Participatory mapping: Exploring landscape values associated with an iconic species. *Applied Geography, 95*, 71–78. https://doi.org/10.1016/j.apgeog.2018.04.013

Gopinath, G. (2010). Critical coastal issues of Sagar Island, east coast of India. *Environmental Monitoring and Assessment, 160*, 555–561. https://doi.org/10.1007/s10661-008-0718-3

Gopinath, G., & Seralathan, P. (2005). Rapid erosion of the coast of Sagar Island, West Bengal – India. *Environmental Geology, 48*, 1058–1067. https://doi.org/10.1007/s00254-005-0044-9

Haklay, M., & Francis, L. (2018). Participatory GIS and community-based citizen science for environmental justice action. In J. Chakraborty, G. Walker, & R. Holifield (Eds.), *The Routledge handbook of environmental justice* (pp. 297–308). Routledge. https://discovery.ucl.ac.uk/id/eprint/1575418/1/24%20Participatory%20GIS%20Haklay%20and%20Francis.pdf

Hedelin, B., Evers, M., Alkan-Olsson, J., & Jonsson, A. (2017). Participatory modelling for sustainable development: Key issues derived from five cases of natural resource and disaster risk management. *Environmental Science & Policy, 76*, 185–196. https://doi.org/10.1016/j.envsci.2017.07.001

Hermans, T. D. G., Šakić Trogrlić, R., van den Homberg, M. J. C., et al. (2022). Exploring the integration of local and scientific knowledge in early warning systems for disaster risk reduction: A review. *Natural Hazards, 114*, 1125–1152. https://doi.org/10.1007/s11069-022-05468-8

Ismail, A., Affriani, A. R., Himayah, S., & Malik, Y. (2019). Participatory mapping for community-based watershed management, lesson learn from central java and west nusa tenggara. In *IOP Conference Series: Earth and Environmental Science* (Vol. 286, No. 1). IOP Publishing. https://doi.org/012024.10.1088/1755-1315/286/1/012024

Kallioras, A., Pliakas, F., Diamantis, I., & Emmanouil, M. (2006). Application of Geographical Information Systems (GIS) for the management of coastal aquifers subjected to seawater intrusion. *Journal of Environmental Science and Health Part A, 41*(9), 2027–2044. https://doi.org/10.1080/10934520600780669

Kettle, N. P., Dow, K., Tuler, S., Webler, T., Whitehead, J., & Miller, K. M. (2014). Integrating scientific and local knowledge to inform risk-based management approaches for climate adaptation. *Climate Risk Management, 4*, 17–31. https://doi.org/10.1016/j.crm.2014.07.001

Liu, W., Dugar, S., McCallum, I., Thapa, G., See, L., Khadka, P., et al. (2018). Integrated participatory and collaborative risk mapping for enhancing disaster resilience. *ISPRS International Journal of Geo-Information, 7*(2), 68. https://doi.org/10.3390/ijgi7020068

Manda, L. T., Notenbaert, A. M. O., & Groot, J. C. J. (2019). A participatory approach to assessing the climate-smartness of agricultural interventions: The Lushoto case. In T. Rosenstock, A. Nowak, & E. Girvetz (Eds.), *The climate-smart agriculture papers*. Springer. https://doi.org/10.1007/978-3-319-92798-5_14

Mere-Roncal, C., Cardoso Carrero, G., Chavez, A. B., Almeyda Zambrano, A. M., Loiselle, B., Veluk Gutierrez, F., et al. (2021). Participatory mapping for strengthening environmental governance on socio-ecological impacts of infrastructure in the Amazon: Lessons to improve tools and strategies. *Sustainability, 13*(24), 14048. https://doi.org/10.3390/su132414048

Mondal, M., Paul, S., Bhattacharya, S., & Biswas, A. (2020). Micro-level assessment of rural societal vulnerability of coastal regions: An insight into Sagar Island, West Bengal, India. *Asia-Pacific Journal of Rural Development, 30*(1–2), 55–88. https://doi.org/10.1177/1018529120946230

Panek, J., & Van Heerden, S. (2013). Participatory GIS for water provision and community planning: Case study Koffiekraal, South Africa. *13th International Multidisciplinary Scientific GeoConference, 1*, 845–851. https://doi.org/10.5593/SGEM2013/BB2.V1/S11.030

Paul, S., Mishra, M., Pati, S., Acharyya, T., Santos, C. A. G., da Silva, R. M., et al. (2024). Evaluation of overwash vulnerability and shoreline dynamics in cyclone-prone Sagar Island, Sundarbans (India). *Science of the Total Environment, 907*, 167933. https://doi.org/10.1016/j.scitotenv.2023.167933

Piñeiro-Corbeira, C., Barrientos, S., Barreiro, R., Aswani, S., Pascual-Fernández, J. J., & De la Cruz-Modino, R. (2022). Can local knowledge of small-scale fishers be used to monitor and assess changes in marine ecosystems in a european context? In I. Misiune, D. Depellegrin, & L. Egarter Vigl (Eds.), *Human-nature interactions*. Springer. https://doi.org/10.1007/978-3-031-01980-7_24

Rahman, A. (2010). *Application of GIS in ecotourism development: A case study in Sundarbans, Bangladesh*. Thesis submitted to the Department of Social Science, Mid-Sweden University, Sweden. https://www.diva-portal.org/smash/get/diva2:326461/FULLTEXT01.pdf

Rawat, P., & Yusuf, J.-E. W. (2020). Participatory mapping, e-participation, and e-governance: Applications in environmental policy. In N. V. Mali (Ed.), *Leveraging digital innovation for governance, public administration and citizen services: Emerging research and opportunities* (pp. 147–175). IGI Global. https://doi.org/10.4018/978-1-5225-5412-7.ch007

Reichel, C., & Frömming, U. U. (2014). Participatory mapping of local disaster risk reduction knowledge: An example from Switzerland. *International Journal of Disaster Risk Science, 5*, 41–54. https://doi.org/10.1007/s13753-014-0013-6

Sanogo, D., Ndour, B. Y., Sall, M., Toure, K., Diop, M., Camara, B. A., et al. (2017). Participatory diagnosis and development of climate change adaptive capacity in the groundnut basin of Senegal: Building a climate-smart village model. *Agriculture & Food Security, 6*(1), 1–12. https://doi.org/10.1186/s40066-017-0091-y

Scully-Engelmeyer, K. M., Granek, E. F., Nielsen-Pincus, M., & Brown, G. (2021). Participatory GIS mapping highlights indirect use and existence values of coastal resources and marine conservation areas. *Ecosystem Services, 50*, 101301. https://doi.org/10.1016/j.ecoser.2021.101301

Tomaquin, R. D. (2023). Best participatory practices in Mangrove conservation management: The case in the Mangrove rehabilitation program in the fishing villages in The Philippines. *Journal of Asian Multicultural Research for Social Sciences Study, 4*(1), 27–31. https://doi.org/10.47616/jamrsss.v4i1.338

Tourinho, L., de Brito Alves, S. M., da Silva, F. B. L., Verdi, M., Roque, N., Conceição, A. A., et al. (2023). A participatory approach to map strategic areas for conservation and restoration at a regional scale. *Perspectives in Ecology and Conservation, 21*(1), 52–61. https://doi.org/10.1016/j.pecon.2022.11.001

UNESCO. (1997). *The Sundarbans*. UNESCO (World Heritage Convention). Retrieved 01 January 2024. Available at: https://whc.unesco.org/en/list/798/

Valbo-Jørgensen, J., & Poulsen, A. F. (2000). Using local knowledge as a research tool in the study of river fish biology: Experiences from the Mekong. *Environment, Development and Sustainability, 2*, 253–376. https://doi.org/10.1023/A:1011418225338

Van Tuan, P., Zhou, Y., Stigter, T., Van Tuc, D., Hai, D. H., & Vuong, B. T. (2023). Design of preliminary groundwater monitoring networks for the coastal Tra Vinh province in Mekong Delta. *Vietnam. Journal of Hydrology: Regional Studies, 47*, 101425. https://doi.org/10.1016/j.ejrh.2023.101425

Yasin, K. H., & Woldemariam, G. W. (2023). GIS-based ecotourism potentiality mapping in the East Hararghe Zone, Ethiopia. *Heliyon, 9*(8), e18567. https://doi.org/10.1016/j.heliyon.2023.e18567

Yu, R., & Mu, Q. (2023). Integration of indigenous and local knowledge in policy and practice of nature-based solutions in China: Progress and highlights. *Sustainability, 15*(14), 11104. https://doi.org/10.3390/su151411104

Zidny, R., Sjöström, J., & Eilks, I. (2020). A multi-perspective reflection on how indigenous knowledge and related ideas can improve science education for sustainability. *Science & Education, 29*, 145–185. https://doi.org/10.1007/s11191-019-00100-x

Further Reading

Albrecht, M. A. (2023). Norwegian seaweed utopia? Governmental narratives of coastal communities, upscaling, and the industrial conquering of ocean spaces. *Maritime Studies, 22*, 37. https://doi.org/10.1007/s40152-023-00324-2

Caquard, S., & Cartwright, W. (2014). Narrative cartography: From mapping stories to the narrative of maps and mapping. *The Cartographic Journal, 51*(2), 101–106. https://doi.org/10.1179/0008704114Z.000000000130

Egberts, L., & Hundstad, D. (2019). Coastal heritage in touristic regional identity narratives: A comparison between the Norwegian region Sørlandet and the Dutch Wadden Sea area. *International Journal of Heritage Studies, 25*(10), 1073–1087. https://doi.org/10.1080/13527258.2019.1570310

Kwan, M.-P. (2008). From oral histories to visual narratives: Re-presenting the post-September 11 experiences of the Muslim women in the USA. *Social and Cultural Geography, 9*(6), 653–669. https://doi.org/10.1080/14649360802292462

Lan, T., O'Brien, O., Cheshire, J., et al. (2022). From data to narratives: Scrutinising the spatial dimensions of social and cultural phenomena through lenses of interactive web mapping. *Journal of Geovisualization and Spatial Analysis, 6*, 22. https://doi.org/10.1007/s41651-022-00117-x

Roth, R. E. (2021). Cartographic design as visual storytelling: Synthesis and review of map-based narratives, genres, and tropes. *The Cartographic Journal, 58*(1), 83–114. https://doi.org/10.1080/00087041.2019.1633103

Segel, E., & Heer, J. (2010). Narrative visualization: Telling stories with data. *IEEE Transactions on Visualization and Computer Graphics, 16*(6), 1139–1148. https://doi.org/10.1109/TVCG.2010.179

Waterman, R. E., Misdorp, R., & Mol, A. (1998). Interactions between water and land in The Netherlands. *Journal of Coastal Conservation, 4*, 115–126. https://doi.org/10.1007/BF02806503

Zhang, Y., & Cheng, Q. (Eds.). (2022). *Geographic information systems and applications in coastal studies*. IntechOpen. https://doi.org/10.5772/intechopen.97909. ISBN: 978-1-80355-742-7.

Chapter 9
Mapping Community Voices in the Coastal Region of Bengal: Case Studies and Best Practices of Participatory GIS

> *In participatory GIS, each dot on the map represents a voice, a story, or a point of view that has to be heard and understood. And maps not only represent reality, but they also shape and influence our perception of it.*
>
> Unknown

Abstract This chapter goes into specific case studies and best practices for utilizing PGIS to map coastal community voices in the Sundarbans. As a consequence, the primary objectives were to map indigenous technical knowledge for fish catching, social sensitivity to climate, participatory mapping of fishing grounds, natural resource management using participatory GIS, and modeling vegetation ecologies in relation to climate change. This chapter used PGIS to involve residents in mapping their surroundings and issues, thereby promoting geospatial citizenship. PGIS provides active engagement in data collection, analysis, and visualization, which improves community insights. This strategy enables individuals to participate in local decision-making processes, advocate for needs, and work with others. Citizens benefit from mapping infrastructure, environmental risks, and cultural sites to better comprehend their surroundings. PGIS promotes collaboration among community members, government institutions, nongovernmental organizations (NGOs), and researchers, hence facilitating collective action. Furthermore, it improves openness in governance by democratizing access to spatial data and encouraging participatory decision-making.

Keywords Indigenous technical knowledge mapping · Applications of participatory GIS techniques · Participatory mapping · Climate change · Natural resource management · Social vulnerability to climate · Mapping of fishing grounds · Traditional marine fishing community · Indian Sundarbans · Coastal region of Bengal · Case studies

This chapter delves into detailed case studies and best practices for using PGIS techniques to map coastal community voices in the Sundarbans. As the outcome, the key goals were to map indigenous technical knowledge for fish catching, social vulnerability to climate, participatory mapping of fishing grounds, natural resource management using participatory GIS, and modeling vegetation ecologies in relation to climate change.

Key Points of the Chapter
- Understanding the application of participatory GIS techniques.
- Participatory mapping with traditional marine fishing coastal communities.
- Exploring and advocating indigenous voices and knowledge through geospatial citizenship.

9.1   Introduction

GIS applications provide a framework for effective data management and the integration of information from various scales and sources, improving understanding of marine resources and their spatial exploitation. This system was used to demonstrate real GIS applications that can help support decisions based on environmental considerations and plan the spatial organization of marine activities (Kirilenko, 2022; Boeser & Hamylton, 2019; Reddy, 2018; Stelzenmüller et al., 2017; Goodchild & Haining, 2004). The use of participatory GIS approaches has enormous promise in the field of spatial voice mapping, allowing for a more nuanced knowledge of communities' opinions and experiences within their physical surroundings (Fagerholm et al., 2021). Participatory GIS indicates that various views are represented and heard during the mapping process, creating more inclusive decision-making processes and actions. By allowing community members to actively participate in mapping activities, this strategy enables individuals to express spatial priorities, identify essential resources, and highlight areas of concern in their neighborhoods or regions (Brown & Kyttä, 2014; Sieber, 2006).

Furthermore, participatory GIS plays an important role in natural resource management by combining local knowledge and scientific data (Levine & Feinholz, 2015; McCall & Minang, 2005). Communities can employ collaborative mapping initiatives to establish resource use zones, conduct risk assessments, document indigenous technical knowledge, and identify places of environmental value. This spatially explicit information informs sustainable resource management strategies in response to climate change, facilitates resource conflict resolution, and promotes biodiversity conservation, spatial planning, and management (Fagerholm et al., 2019; Eilola et al., 2014; McCall & Dunn, 2012; Termorshuizen & Opdam, 2009). Several researchers working on PGIS applications with a worldwide focus have highlighted community engagement in spatial monitoring and management. In this regard, we investigated the following findings:

Macnab (2002) conducted a research on local fishermen's knowledge and perceptions about fishing and other activities. Kwaku Kyem (2004) did study on conflict resolution and assessment using PGIS in Southern Ghana. In their 2005 study, McCall and Minang explored the use of participatory GIS for natural resource management in Cameroon communities. Ramsey (2008) investigated the politics of geographical knowledge generation by directly juxtaposing opposing understandings of space and spatial challenges through participatory GIS methodologies. Baldwin and Mahon (2014) addressed maritime spatial planning using PGIS in the Grenadines. Levine and Feinholz (2015) explored coral reef ecosystem management using participatory GIS and identified stakeholder engagement in the Hawaiian Islands. Bhendekar et al. (2016) developed thematic maps of fisheries resources from trawl fisheries around the Mumbai coast using PGIS. Rakotomahazo et al. (2019) investigated the participatory design of community-based payments for ecosystem services in Madagascar's mangrove society. Sullivan-Wiley et al. (2019) used participatory community mapping in eastern Uganda to identify vulnerabilities. Fagerholm et al. (2019) undertook a study of participatory spatial planning in Tanzania's Southern Highlands. Abdel-Raheem (2019) studied community-based marine spatial planning using participatory methodologies on St. George's Caye, Belize. Paulangan et al. (2020) researched the fishing season and participatory mapping of fishing grounds in Papua Indonesia. da Silva et al. (2021) investigated several ways to detecting possible conflicts and determining positioning accuracy in marine and coastal ecosystem services. Escandón-Panchana et al. (2022) conducted a case study in Punta Carnero, Ecuador, to study the spatial design of the coastal marine socioecological system. Grati et al. (2022) adopted the Mediterranean Sea as a case study to investigate the mapping of fishing grounds and other aspects of fisheries management via participatory strategies.

Therefore, the majority of these studies have focused on participatory GIS techniques and the incorporation of local shareholder participation in on-the-ground mapping for decision-making and spatial planning. Based on this setting, this chapter thoroughly explores and evaluates a series of case studies for studying Indian coastal ecologies by mapping the voices of coastal geospatial inhabitants. Specifically, the following case studies were considered:

Case study 1 Mapping indigenous technical knowledge of fish catching (the case of coastal Sundarbans)

Case study 2 Mapping the social vulnerability to climate (the case of coastal region of Bengal)

Case study 3 Participatory mapping of fishing grounds (the case of coastal region of Bengal)

Case study 4 Natural resource management using participatory GIS (the case of coastal Sundarbans)

Case study 5 Modeling vegetation ecologies in relation to climate change (the case of coastal South 24 Parganas)

9.2 Exploring Indian Coastal Ecologies: A Series of Case Studies on Mapping the Voices of Geospatial Citizens

Case Study 1

9.2.1 Participatory GIS Tools for Mapping Indigenous Technical Knowledge of Fish Catching in the Traditional Marine Fishing Community: The Case of Coastal Sundarbans

Background and Purpose

Indigenous technical knowledge (ITK) are crucial for fish catching in traditional marine fishing communities (Poto et al., 2022; Edwin et al., 2019; Shyam & Antony, 2013), which is beneficial to regional sustainability (Roy et al., 2020; Shenoy, 2009; Ponnusamy et al., 2009). Utilizing PGIS tools proves highly effective in mapping ITK gathered from local stakeholders (Kitolelei et al., 2022; Lamptey, 2009). Based on this background, we collaborate with the local traditional marine fishing community peoples to introduce and comprehend the various PGIS tool-based knowledge for mapping the ITK of fish catching in the coastal Sundarbans (Fig. 9.1).

This case study was conducted from September to November 2023 using focus group discussions (FGD), Google Maps and Google Earth applications, community mapping workshops, GIS training and capacity building, story mapping, and other methodologies (including sketch mapping and interactive sketch mapping). Twenty-eight (28) locations were visited, and 60 citizens took part in this event.

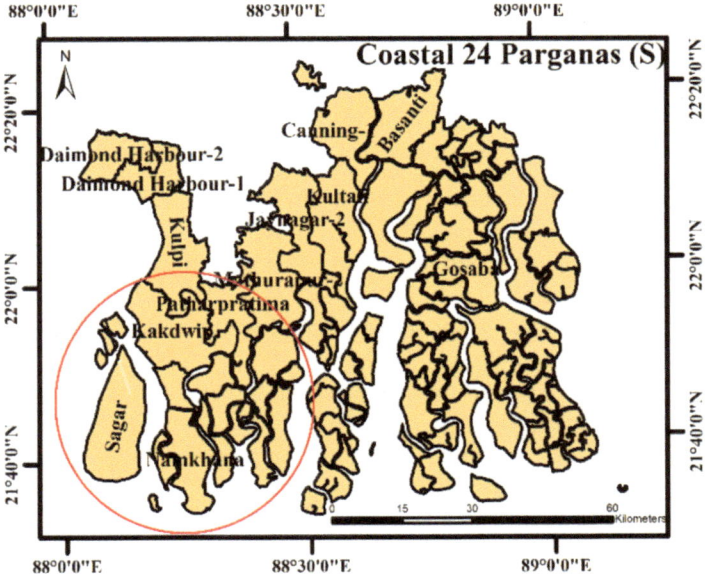

Fig. 9.1 Mapping study area (Coastal Sundarbans: Sagar Island, Kakdwip, Namkhana, and Pathar Pratima)

Results, Discussion, and Conclusion

Table 9.1 shows the application of PGIS methodologies in the Indian Sundarbans, specifically in the Sagar Island, Kakdwip, Namkhana, and Pathar Pratima regions.

Table 9.1 Application of PGIS tools and techniques for mapping ITK

Sr. no.	PGIS tools and techniques	Discussions/themes/applications	Scientific evidence(s)/figure(s)
01	FGD 1 (climate change and declination of fishing species)	*Discussions*: Rising temperatures, less rainfall, increased cyclonic activity, and habitat destruction all pose serious concerns for the survival of fish species in the coastal Sundarbans *Remarks*: Immediate conservation efforts and sustainable management methods are required to offset these effects and protect biodiversity in this fragile coastal ecosystem	Ghosh and Roy (2022), Gopal and Chauhan (2006), and World Bank (2014)
02	FGD 2 (climate change, mangrove losses, and altered fishing habitats)	*Discussions*: Climate change promotes mangrove destruction and affects fishing habitats in the coastal Sundarbans. Rising sea levels, severe weather events, and ocean acidity all contribute to habitat loss *Remarks*: This threatens fish populations and the livelihoods that rely on them. Urgent conservation measures, sustainable fishing methods, and adaptation strategies are critical for protecting both ecosystems and coastal communities	Yamamoto (2023), Salvatteci et al. (2022), and Carugati et al. (2018)
03	FGD 3 (climate change, coastal erosion, cyclonic disasters, and changes of fishing grounds)	*Discussions*: Rising sea levels exacerbate erosion, and greater cyclonic activity disturbs fisheries and habitats. These changes endanger the lives of coastal populations that depend on fishing. *Remarks*: To reduce the effects of climate change and protect the Sundarbans' ecosystems and livelihoods, immediate action is required, such as coastal protection measures, disaster preparedness, and sustainable fishing techniques	Ghosh and Mistri (2023), Muhala et al. (2021), Barange et al. (2018), and Islam et al. (2014)
04	Google maps (location identifications)	During the field investigation, many sites were located within the Sagar Island and Namkhana blocks that served as a "Ghat" (wharf) for fishing. This was validated by locating the drop-off locations with the Google maps mobile app	Figure 9.2
05	Google maps (finding route in land)	Residents of TMFC, especially the younger generation, were accustomed to mapping their insights to identify the routes leading to fishing wharves on land. Primarily, they understood how to navigate these routes and recognized the advantages of traveling from one known location to another unknown destination	Figure 9.3

(continued)

Table 9.1 (continued)

Sr. no.	PGIS tools and techniques	Discussions/themes/applications	Scientific evidence(s)/ figure(s)
06	Share live location through WhatsApp	Utilizing mobile apps like WhatsApp, which are widely popular in India with almost every internet user having used them, serves as an effective PGIS tool. Fishermen in the study area employ WhatsApp for regular conversations, voice and video calls, and real-time location sharing, aiding each other with navigation and facilitating urgent communication while fishing	Figure 9.4
07	Google earth pro (mapping possible fishing grounds)	According to our field investigation, we identified possible fishing areas on Sagar Island, Kakdwip, Namkhana, and Pathar Pratima. We observed both major and minor fishing locations, as well as transforming places. Local fishermen argue that some areas have shifted as a result of sediment deposition and the negative effects of climate change	Figure 9.5
08	Google earth pro (mapping river routes)	With input from local residents, we charted out the safest river routes for fishing and identified specific fishing zones beneficial for marginal fishermen	Figure 9.6
09	Community mapping workshops/GIS training/sketch mapping	During the focus group discussions, we addressed the basics of GIS tools and participatory approaches to community engagement in coastal zone management and fishing zone protection. As well, we mapped the sketch (using community conversations) to gain a better grasp of their mental maps	Figure 9.7
10	Story mapping	During the interview, Mr. A. Hazra[1] (age 66 Yrs.) recalled his family's migratory journey, including storms, coastal floods, saltwater intrusion, food insecurity, and unemployment. Following a comprehensive review of his life path, we defined it as a "climate migration" due to a variety of climate-induced vulnerabilities and other risk factors that underpin this spatial occurrence [1]Mr. A. Hazra: A revered participant from Sagar Island (West Bengal, India), whose name has been altered for confidentiality	Figure 9.8

Note: The author (S. Roy) visited the Indian Sundarbans from September to November 2023 to collect the necessary data

TMFC traditional marine fishing community, *FGD* focus group discussions, *Google maps* using mobile app, *WhatsApp/share live location* using mobile app; Google earth pro (V 7.3.6.9750): Available at https://earth.google.com/

This study mapped the fishing community's daily lives and livelihood expressions in relation to identifying fishing grounds, best possible river and stream routes, socio-ecological knowledge about fishing species decline, transformation of fishing grounds, local sea level rise, and coastal erosion zones in terms of climate change and highlights the delta location. Furthermore, we taught the local community about the applications and uses of PGIS technologies through community mapping workshops, and we emphasized the sketches for their narratives using the ideas of mental maps and story mapping. In conclusion, we pointed out that different types of knowledge mapping are valuable to decision-makers, policy planners, and researchers. These technologies help to identify new research directions, fine-tune policy approaches, and make spatial decisions more efficient. Furthermore, they serve to improve community literacy in PGIS applications, allowing people to become active geospatial citizens.

Case Study 2

9.2.2   Mapping the Social Vulnerability to Climate in the Coastal Community: A Case Study of the Coastal Region of Bengal

Background and Purpose

Climate change is a major source of concern among coastal communities (Hemani, 2014). As sea levels rise and storms worsen, marginalized communities are frequently hit hardest because of factors such as insufficient resources, inadequate infrastructure, and unequal access to information and support networks (Shiiba et al., 2023; Johnson et al., 2023). Vulnerable populations, such as indigenous communities, low-income households, and the elderly, are at a higher risk of displacement, health problems, and financial difficulty (White et al., 2023; Shivanna, 2022; Jayawardhan, 2017). To develop resilience and assure no one is left behind in the face of climate-related issues, social vulnerability must be addressed holistically, with a focus on equality, community participation, and inclusive decision-making (Lv et al., 2024; Carmen et al., 2022; Nunes, 2021; Nutters, 2012).

Based on these contexts, we used a mixed method approach (including local community participation) in the coastal region of Bengal (Fig. 9.9) to map the coastal community's social vulnerability to climate change. In this environment, a case study was undertaken between April and June 2022, highlighting potential social vulnerability facts.

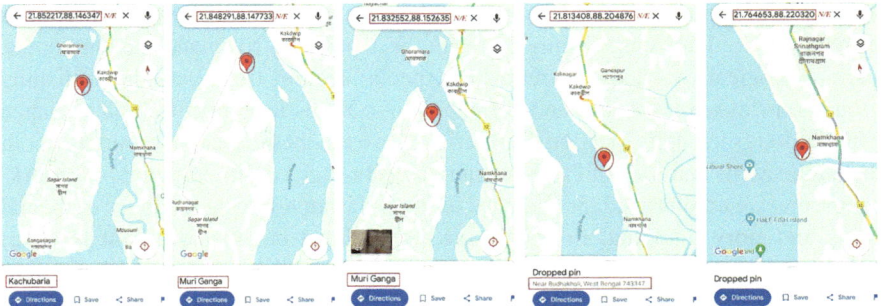

Fig. 9.2 Location identification (fishing wharf) using Google Maps in Sagar Island and Namkhana blocks

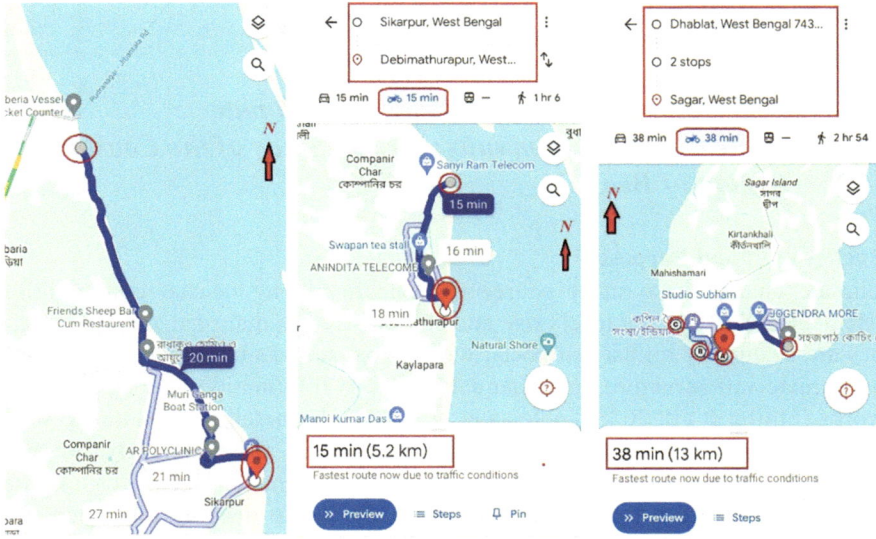

Fig. 9.3 Routes navigated by the traditional marine fishing community in the Sagar Island, West Bengal

Results, Discussion, and Conclusion

(a) *Vulnerability in Fish Catching*

The field survey (Fig. 9.10) revealed that all fishing populations recognized the significant impact of various factors, such as extreme events, monsoon patterns, rainfall variability, sea surface salinity, seasonal fluctuations, water color changes, and cyclone frequency, on their fishing activities and fish availability (Pattanaik, 2021). A majority (70.63%) attributed the extinction of traditional sea species to climate change and related hazards, while 34.79% reported the emergence of new species. Overall, respondents expressed concern about the vulnerability of their livelihoods to climate-related influences.

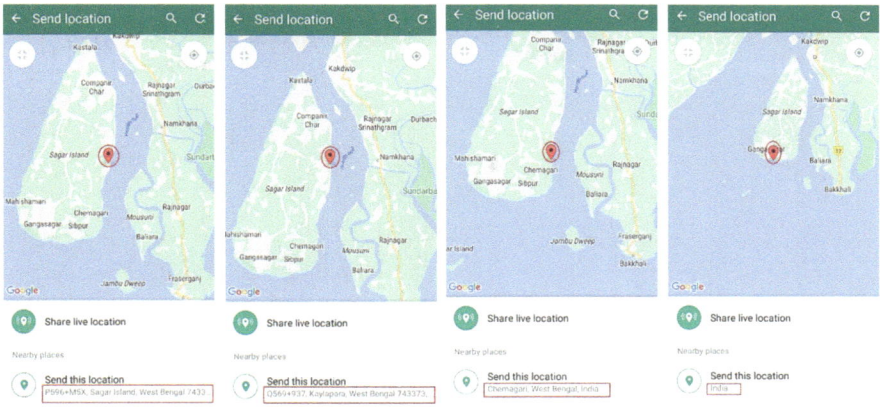

Fig. 9.4 Shared the "live location" on Sagar Island via WhatsApp

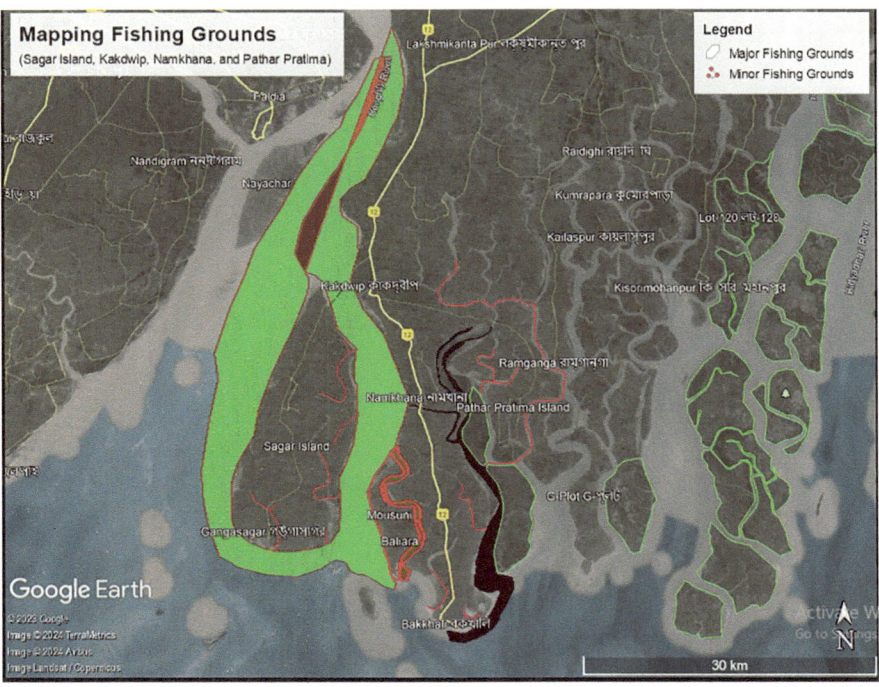

Fig. 9.5 Mapping possible fishing grounds in the Sagar Island, Kakdwip, Namkhana, and Pathar Pratima (West Bengal)

(b) *Vulnerability and Migration*

During the FGDs, community populations were identified; some of the low-lying Sundarban communities moved from one location to another (Sagar Island) as a result of the ongoing impact of storms, coastal floods, saltwater

Fig. 9.6 Mapping river routes and marginal fishing zones

intrusion, loss of daily livelihoods (fish catching), food insecurity, and unemployment (Dasgupta, 2018). Figure 9.8 already shows a sketch mapping of climate migration using community mapping, highlighting climate vulnerability scenarios.

(c) *Vulnerability and Climate Gap*

Based on the qualitative investigation, we observed inequities in climate change impacts. Vulnerability variables such as socioeconomic status and location influence risk. Marginalized populations on Sagar Island bear disproportionate hardships due to inadequate resources and increased vulnerability to risks (Fig. 9.11) (Paul et al., 2024). In this regard, we propose that closing the gap entails addressing systemic injustices, promoting equity, and strengthening resilience through inclusive policies and decision-making, as well as implementing participatory community engagement and the development of spatial information among local citizens.

(d) *Vulnerability and Climate Poverty*

Based on the observation (climate migration and climate gaps in coastal populations), people who lack resources and endure exposure to environmental stressors were more vulnerable to climate-related disasters and disruptions, highlighting the cycle in which environmental degradation leads to economic hardships, further deepening vulnerability, and perpetuating poverty.

Fig. 9.7 Community mapping: a journey of Cyclone Aila (May 25–27, 2009)

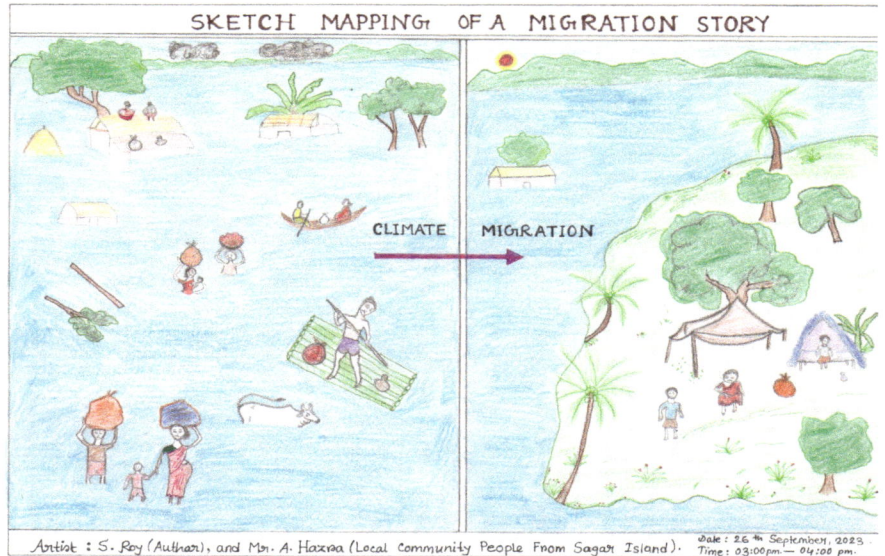

Fig. 9.8 Sketch mapping of a migration story

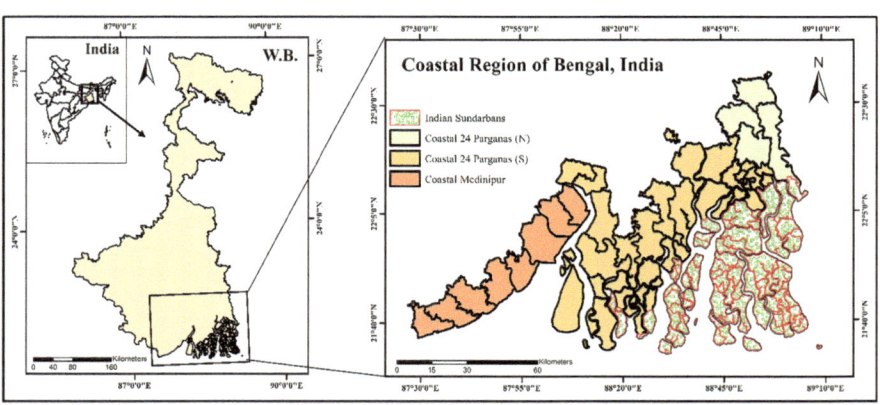

Fig. 9.9 Mapping study area (coastal region of Bengal, India)

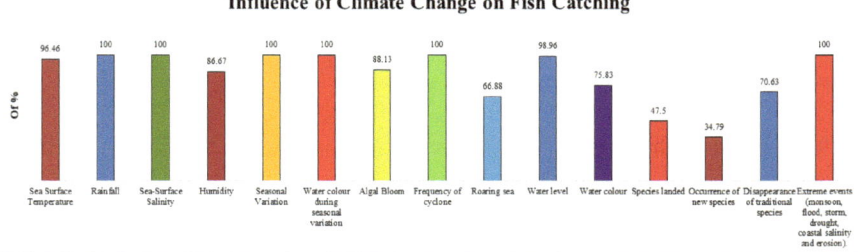

Fig. 9.10 Influence of climate change on fish catching in the coastal region of Bengal, India

Fig. 9.11 Lack of proper resources and infrastructures in the rural Sagar Island, West Bengal. (Source: Primary field survey by the Author (K.D. Malakar))

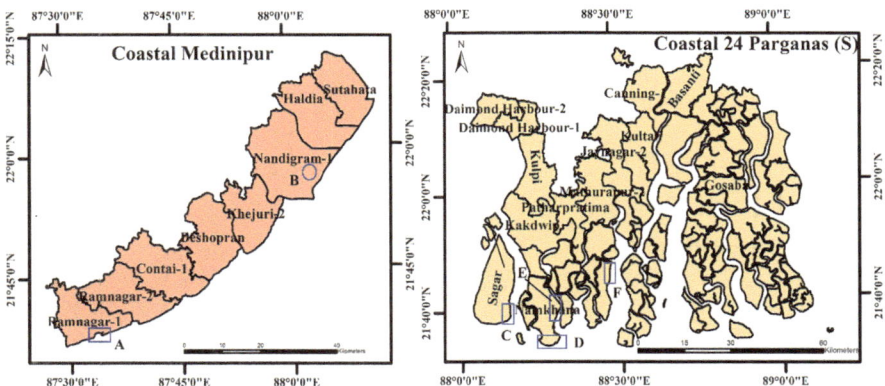

Fig. 9.12 Participatory mapping of climate risk in agriculture and aquaculture. (**a**) Climate risk on aquaculture land; (**b**) climate risk on agricultural land; and (**c–f**) Climate risk on agriculture and aquaculture transformed land

(e) *Climate Risk on Agriculture and Aquaculture*

During the community mapping workshops, people from the fishing community identified (Fig. 9.12) vulnerable and transformed land usage during the last 10 to 20 years. They located (using Google Earth Pro Mapping) sensitive agricultural area and transformed agricultural land (Fig. 9.13) as a result of climate change (Priyadarshini, 2015).

Fig. 9.13 Mapping the vulnerable cyclone-affected areas

(f) *Impact of Climate Change on Human Health*

In the 2022 field survey in Bengal's coastal area, the traditional marine fishing community faces various health challenges due to climate change. Figure 9.14 shows data on infectious, water-borne, and vector-borne diseases; child health issues; and mortality rates. Respondents noted a worsening impact of climate change on health over the past 5 years, with no improvements. Engaged mainly in fishing, community members are regularly exposed to water-related activities. Additionally, issues like food insecurity and poverty worsen these health concerns. Overall, climate change and existing socioeconomic challenges compound health issues within this community.

(g) *Climate Disasters, Political Ambivalence in Relief Distribution, Local Politics, Illegal Immigration, and Societal Transformation*

According to field data collected in 2022, 78.33% of respondents in the Indian Sundarbans stated experiencing political ambiguity during aid distribution. Local politics influenced the uneven setup of disaster aid management. Some people were mentioned regarding illegal immigration from Bangladesh, where migrants frequently lack identification and legal status, live in terrible conditions, and rely on illegal activities such as theft and smuggling to survive.

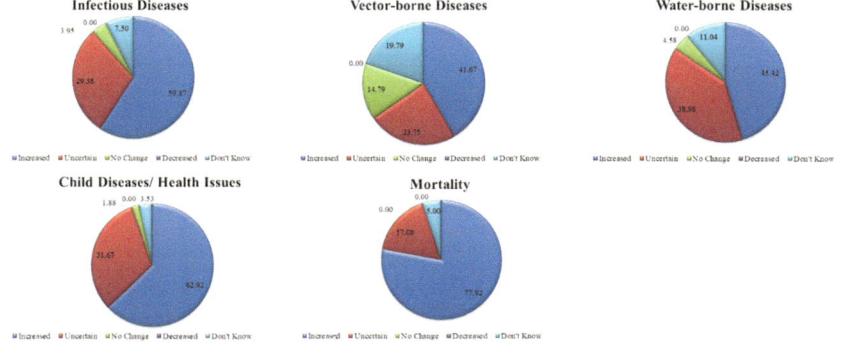

Fig. 9.14 Climate-induced health problems (infectious, water-borne, and vector-borne diseases, child health issues, and mortality)

Fig. 9.15 Possible paths of illegal immigration in the coastal Sundarbans

Although many immigrants are unvaccinated, there has been an increase in crime and disease spread in the area, which has served as an additional driver for societal transformation. In this context, the local community highlighted (Fig. 9.15) the potential ways for illegal immigration, which was another instance of social transformation.

Hence, this case study focuses primarily on delineating the spatial vulnerability of the study area through the active participation of local communities. The key takeaway from this study underscores the necessity of integrating local citizens into the development of resilience frameworks and policy planning, emphasizing the inherent spatial dimension of sustainability that must be upheld.

Case Study 3

9.2.3   Participatory Mapping of Fishing Grounds in the Traditional Marine Fishing Community in the Coastal Region of Bengal

Background and Purpose
The identification of spatial fishing grounds with the assistance of traditional fishermen is useful for fisheries planning and management (André et al., 2022; Said & Trouillet, 2020). This planning and management will assist us in preserving local biodiversity, managing coastal zones, conserving marginal fishing species, and other socio-spatial developments (Dineshbabu et al., 2019; Janßen et al., 2018). This case study gathered GPS location information for traditional fishing areas in the coastal region of Bengal, India (Fig. 9.16). Between September and November 2023, over 120 community populations were engaged in data collection, and Google Earth Pro, an outstanding PGIS tool, was utilized to plot these points and emphasize the fishing grounds.

Results, Discussion, and Conclusion
Based on the GPS data collected in the field (2023), we tubulated (Table 9.2) the locational coordinates of the traditional fishing grounds and plotted them (Fig. 9.17)

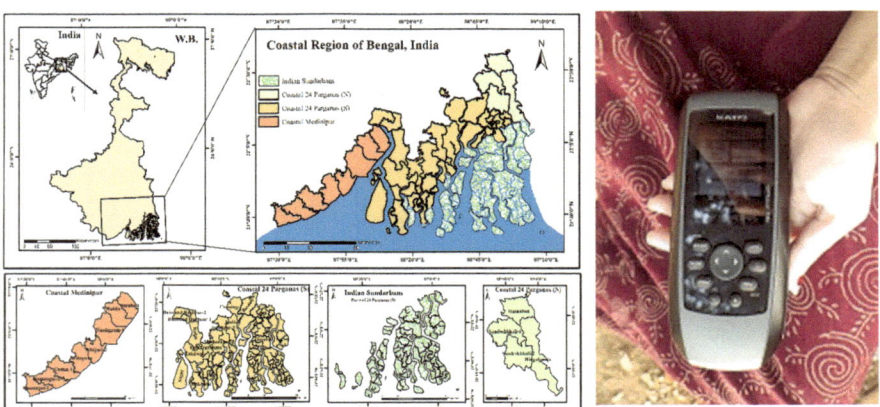

Fig. 9.16 Mapping the study area (coastal region of Bengal, India). And used GPS instrument during the field visit

Table 9.2 GPS-based identification of traditional fishing sites in the coastal region of Bengal (India)

Sr. no.	Fishing grounds (FG)	Latitude	Longitude	Sr. no.	Fishing grounds (FG)	Latitude	Longitude
1	FG 1	21°37′31.03″N	87°32′2.06″E	22	FG 22	21°48′2.29″N	88°26′39.44″E
2	FG 2	21°38′11.12″N	87°33′14.35″E	23	FG 23	21°51′55.59″N	88°32′24.57″E
3	FG 3	21°38′16.94″N	87°33′48.96″E	24	FG 24	22°10′12.19″N	88°52′12.97″E
4	FG 4	21°38′36.39″N	87°37′43.57″E	25	FG 25	22°21′43.87″N	88°51′10.45″E
5	FG 5	21°40′4.28″N	87°38′17.89″E	26	FG 26	22°22′23.31″N	88°53′1.17″E
6	FG 6	21°39′37.48″N	87°38′49.78″E	27	FG 27	22°13′22.05″N	88°57′15.98″E
7	FG 7	21°47′55.17″N	87°54′40.43″E	28	FG 28	22°15′26.17″N	88°53′25.35″E
8	FG 8	21°49′4.67″N	87°56′12.90″E	29	FG 29	22°12′50.44″N	88°39′19.20″E
9	FG 9	21°56′23.31″N	88° 1′24.05″E	30	FG 30	22°12′15.41″N	88°42′35.49″E
10	FG 10	22° 1′18.41″N	88° 2′35.17″E	31	FG 31	22°11′0.41″N	88°46′5.53″E
11	FG 11	21°52′14.74″N	88° 8′39.61″E	32	FG 32	22° 8′28.68″N	88°47′56.05″E
12	FG 12	21°50′20.39″N	88° 9′6.74″E	33	FG 33	22° 5′23.77″N	88°45′47.84″E
13	FG 13	21°48′33.40″N	88°10′18.98″E	34	FG 34	22° 52.74″N	88°45′8.89″E
14	FG 14	21°41′3.75″N	88° 9′13.82″E	35	FG 35	22° 0′22.14″N	88°43′11.13″E
15	FG 15	21°38′48.48″N	88° 9′10.84″E	36	FG 36	22° 0′32.73″N	88°42′39.17″E
16	FG 16	21°47′12.71″N	88° 4′58.83″E	37	FG 37	22° 0′37.76″N	88°41′56.07″E
17	FG 17	21°41′55.94″N	88° 2′53.03″E				
18	FG 18	21°39′51.75″N	88°13′47.88″E				
19	FG 19	21°43′50.98″N	88°15′42.91″E				
20	FG 20	21°45′26.30″N	88°14′53.74″E				
21	FG 21	21°48′42.25″N	88°22′9.26″E				

Source: Primary field survey (2023)

N.B. Data collected by the author (S. Roy)

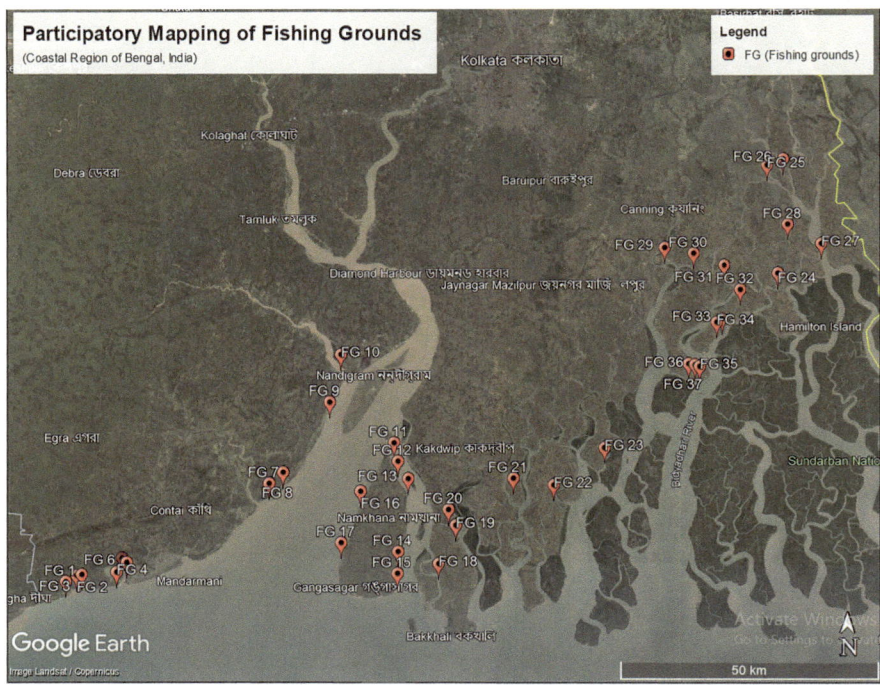

Fig. 9.17 Participatory mapping of traditional fishing grounds in the coastal region of Bengal, India

for further comprehension. We extend the PGIS approach to include sustainable fishing practices and participatory decision-making for spatial management. Therefore, involving the traditional marine fishing population in Bengal's coastal region through participatory mapping of fishing grounds is critical. This collaborative method allows community members to contribute local knowledge, assuring accurate depiction of resource distribution and consumption, and sharing their expertise of climate-related fisheries challenges. By encouraging active participation, it fosters a sense of ownership over marine resources and supports sustainable fishing methods. This effort lays the groundwork for effective fisheries management by using participatory decision-making to balance conservation goals with the community's socioeconomic requirements. In the end, participatory mapping acts as a stimulus for community-driven solutions and the long-term sustainability of coastal livelihoods.

Case Study 4

9.2.4 Mapping the Scope of Natural Resource Management Using Participatory GIS: A Case Study of the Indian Sundarbans

Background and Purpose

Natural resource management utilizing participatory GIS leverages community interaction and spatial analysis to inform long-term practices. This strategy assures an accurate depiction of resource dynamics and community requirements by including local stakeholders in data collection and decision-making (Fagerholm et al., 2021). Integrating traditional knowledge with modern technologies makes it easier to establish specialized management methods that balance ecological preservation and socioeconomic development (Andrade-Sánchez et al., 2021; Tripathi & Bhattarya, 2004).

In this regard, we organized a community workshop to map the natural resource management scopes (local mangrove forests) with community participation from the Indian Sundarbans (Fig. 9.18). During the field survey, which runs from September to November 2023, 30 community populations are involved in this conversation. During the discussion, we focused mostly on local community engagement and the integration of traditional knowledge with cutting-edge technology.

Result, Discussion, and Conclusion

Figure 9.19 illustrates a framework for incorporating the local community into mapping natural resource management, which is applicable to participatory policy planning in any forest management context. Thus, blending local citizens' traditional knowledge and technical assumptions with scientific planning will enhance spatial management and promote sustainability.

Case Study 5

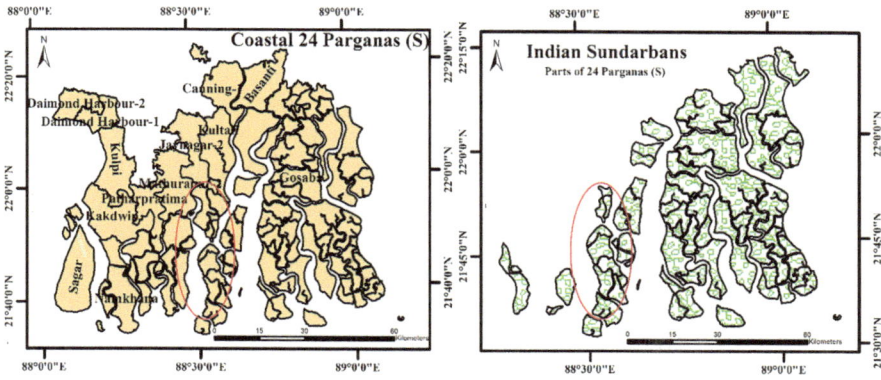

Fig. 9.18 Mapping study area (Indian Sundarbans)

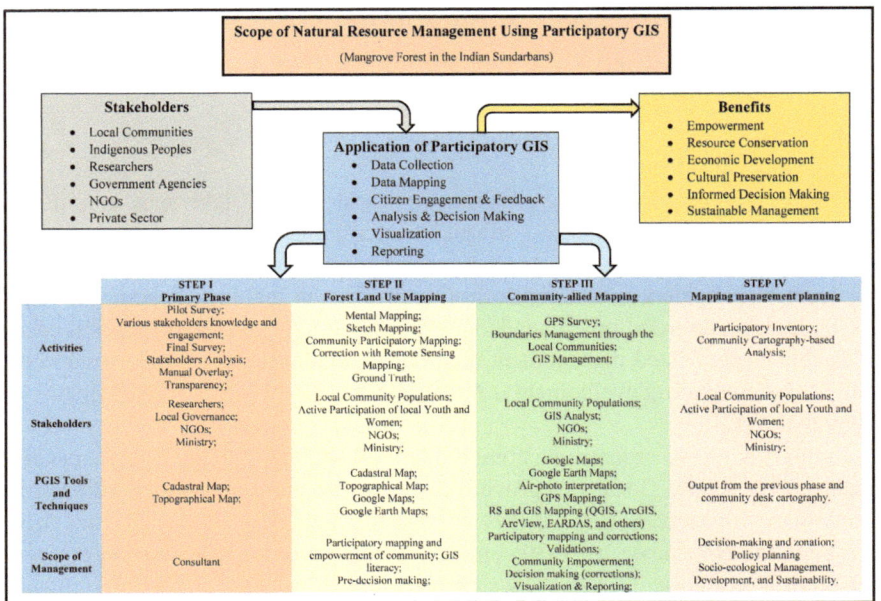

Fig. 9.19 Scope of natural resource management using PGIS. (Made by the authors)

9.2.5 Modeling Vegetation Ecologies in Relation to Climate Change, Using Remote Sensing and PGIS

Background and Purpose

Coastal vegetation ecologies (including mangroves) are vital ecosystems that provide several social, economic, and ecological benefits, such as carbon sequestration, coastal protection, and habitat for a variety of marine species (Madhav et al., 2022; Sandilyan & Kathiresan, 2012; Acharya, 2002). However, these essential ecosystems are becoming increasingly vulnerable to the effects of climate change, including as variations in precipitation patterns, rising temperatures, sea level rise, and extreme weather events (Akram et al., 2023; Malakar, 2020; Mitra, 2013). To properly understand and address these risks, cutting-edge modeling techniques combining remote sensing and participatory GIS need to be used. These technologies allow for the long-term monitoring of changes in forest area, structure, and health, which provides vital insights into the effects of climate change (Massey et al., 2023; Gabriele et al., 2023). On the other hand, participatory GIS brings together local communities, stakeholders, and professionals to combine traditional ecological knowledge with scientific data to improve model accuracy and relevance, encouraging a comprehensive understanding of vegetation ecosystems and their reactions to climate change (Diswandi, 2022; Dawit & Simane, 2017; Keenan, 2015).

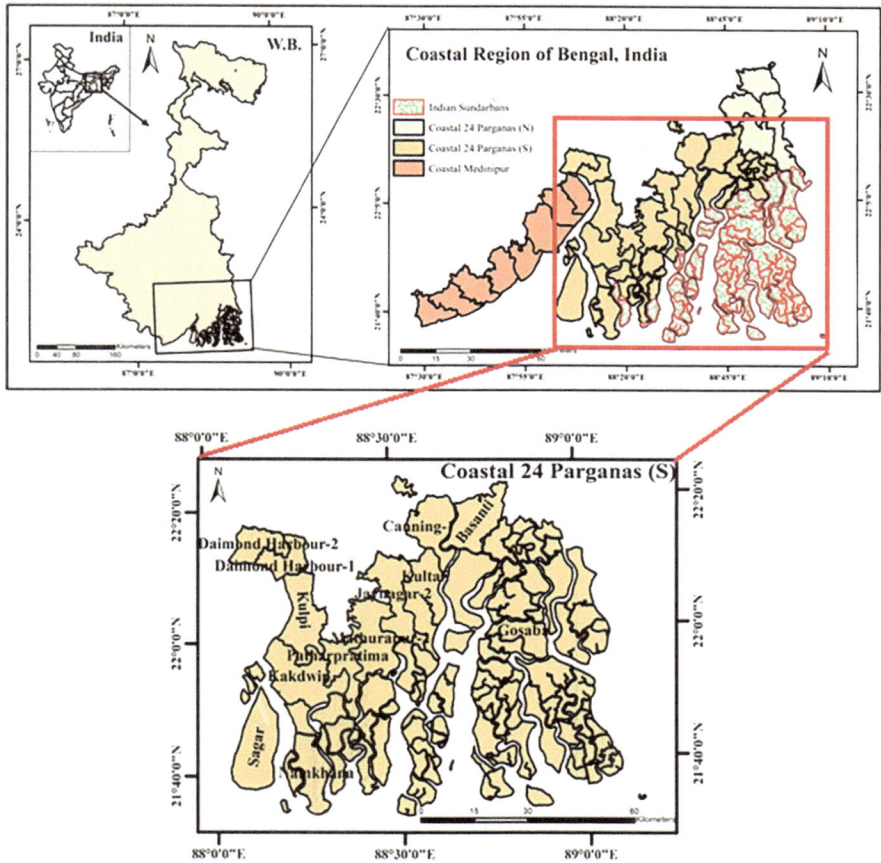

Fig. 9.20 Study area mapping (coastal South 24 Parganas, West Bengal, India)

Based on this background, we employed remote sensing-based vegetation index approaches, normalized difference vegetation index (NDVI) Eq. 9.1 (Pandya et al., 2023; Gandhi et al., 2015), and the supervised classification techniques (Jana et al., 2016) to monitor the seasonal (pre- and post-monsoon) NDVI mapping and vegetation covers of the Coastal 24 Parganas (S) (Fig. 9.20) in 1990, 1998, 2005, 2010, 2015, and 2022. Furthermore, we investigated the seasonal variation of climate variables during the previous 32 years (1990–2022), as well as the impact and interaction of these changes on NDVI and vegetation cover in the study area. In this scenario, we obtained high-resolution gridded climate datasets (CRU TS version 4.07l source: https://crudata.uea.ac.uk/cru/data/hrg/cru_ts_4.07) and Landsat satellite data from USGS Earth Explorer (source: https://earthexplorer.usgs.gov). Furthermore, we did a field inquiry (N = 300) during April to June 2022 and utilized participatory approaches to highlight the most important factor influencing vegetation health risks in the study region by using Garrett ranking techniques (Eq. 9.2) (Garrett, 1979).

$$NDVI = \frac{(NIR - RED)}{(NIR + RED)} \tag{9.1}$$

where NIR is the reflectance value in the near-infrared band and RED is the reflectance value in the red band.

$$Percent\ Positions = \frac{100(Rij - 0.5)}{Nj} \tag{9.2}$$

where "R_{ij}" is the rank given for the ith factors by the jth respondents and "Nj" is the number of factors ranked by the jth respondents.

Result, Discussion, and Conclusion

To highlight the impact of climate change and variability, we analyzed 32 years of climatic data encompassing both pre- and post-monsoon seasons. Our study revealed significant variability-driven changes in the study area, as illustrated in Fig. 9.21.

Pre-Monsoon Season

From 1990 to 2022, Table 9.3 provides pre-monsoon seasonal climatic data (temperature and precipitation), NDVI (Fig. 9.22), and vegetation cover (Fig. 9.23). The

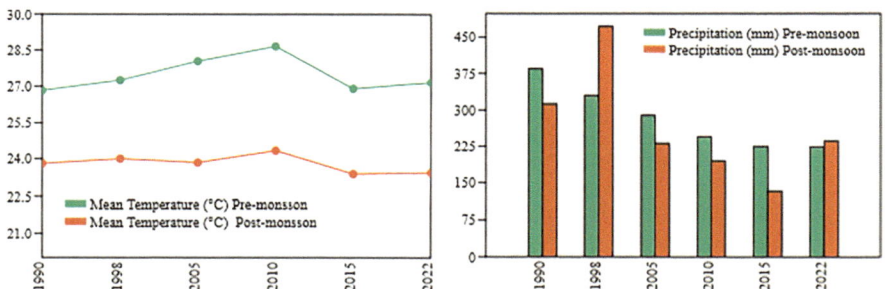

Fig. 9.21 Season-wise variation of climatic parameters (temperature and precipitation) in coastal 24 Parganas (S) (1990–2022)

Table 9.3 Values of NDVI, vegetation cover, and climatic parameters of coastal 24 Parganas (S) (pre-monsoon season)

Year	NDVI	Vegetation cover (sq. km)	Mean temperature (°C)	Precipitation (mm)
1990	0.072	2069.441	26.834	384.531
1998	0.046	2283.166	27.247	329.749
2005	0.079	2215.456	28.033	289.482
2010	0.044	2199.407	28.649	245.379
2015	0.162	1938.572	26.898	225.935
2022	0.179	1414.354	27.157	224.743

Source: Calculated by the author

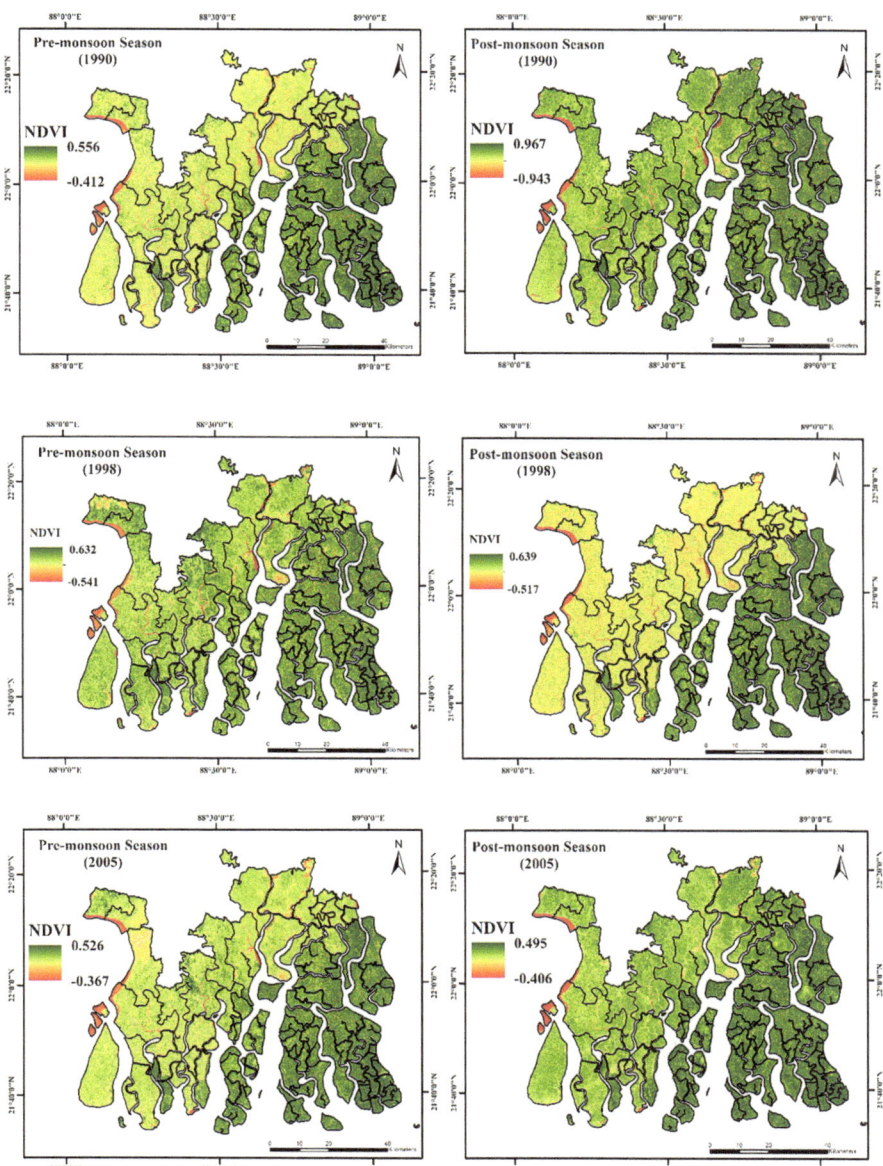

Fig. 9.22 Seasonal (pre-monsoon and post-monsoon) NDVI mapping of the coastal 24 Parganas (S) during 1990, 1998, 2005, 2010, 2015, and 2022

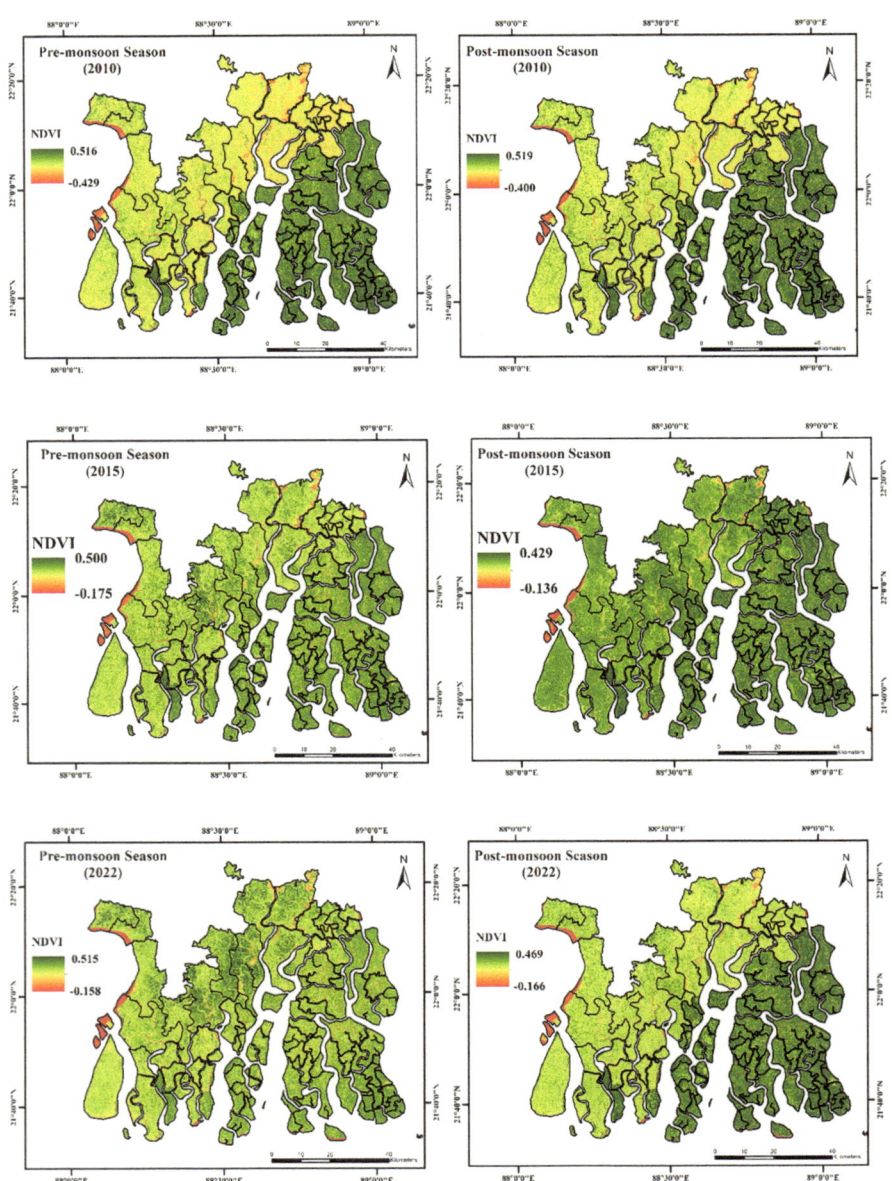

Fig. 9.22 (continued)

link between these elements and their fluctuations was observed, and Figs. 9.24 and 9.25 depict the relationship between climatic variables (temperature and precipitation), vegetation variables, and vegetation covering. Figure 9.24 illustrates a cartographical representation of the relationship between climatic variables and vegetation, as indicated by NDVI. Temperature and NDVI have a moderate to low regression coefficient of 0.2615, whereas precipitation and NDVI have a moderate

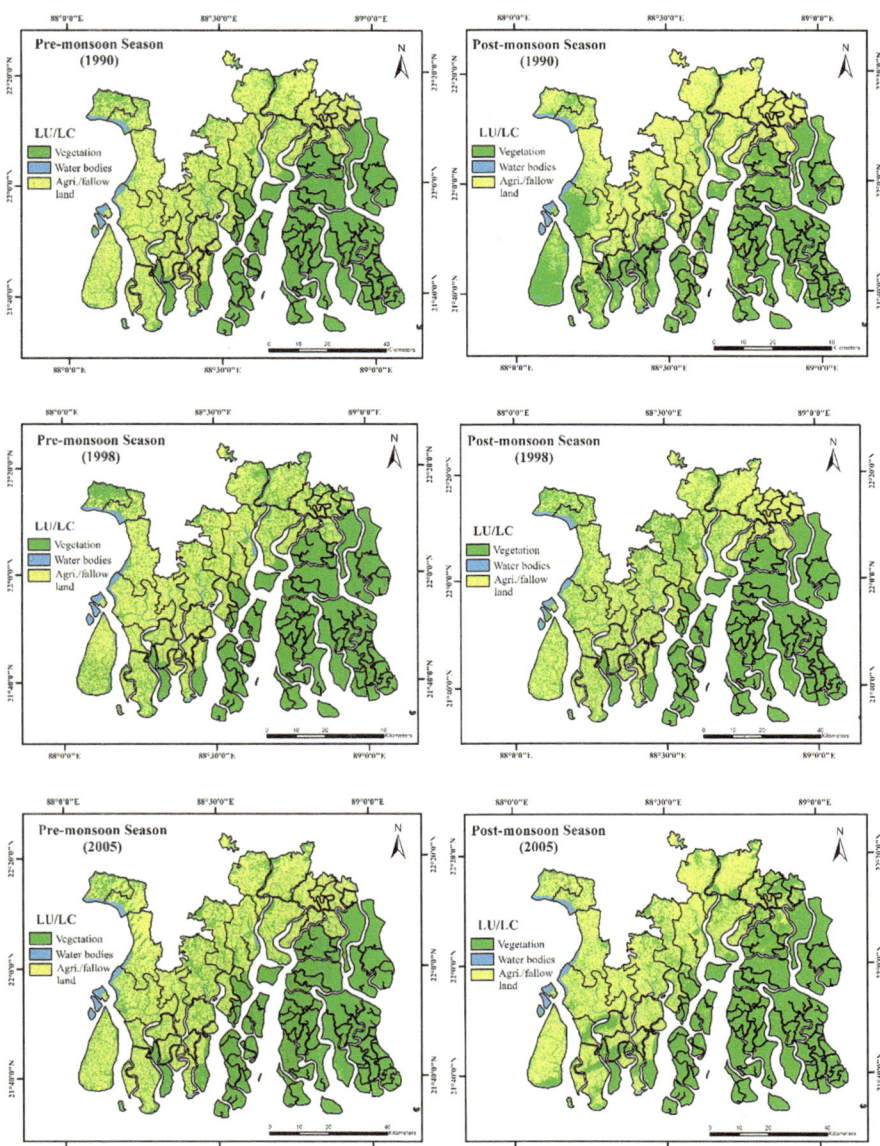

Fig. 9.23 Seasonal (pre-monsoon and post-monsoon) LU/LC mapping of the coastal 24 Parganas (S) during 1990, 1998, 2005, 2010, 2015, and 2022

to low regression coefficient of 0.3730. Similarly, regression analyses of temperature, precipitation, and vegetation coverage provide moderate to low regression coefficient values of 0.1640, 0.2524, and 0.2524, respectively.

Overall, the study's findings indicate a significant (moderate to low) relationship between altering climatic variability conditions during the study area's pre-monsoon season. In terms of precipitation variability, NDVI and vegetation cover have higher

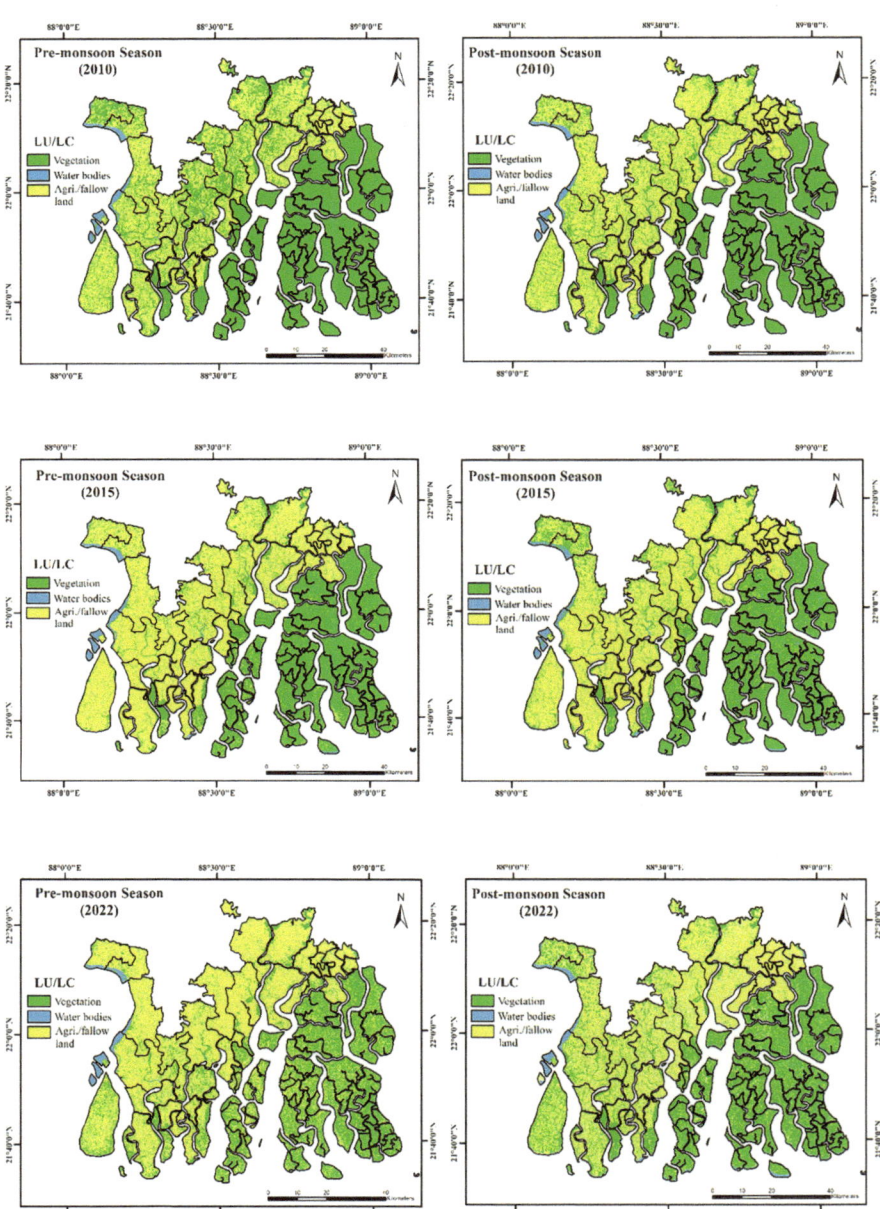

Fig. 9.23 (continued)

coefficient values than temperature variability. Water provides nutrients to plants and trees throughout the season, but other biological activities associated with vegetation growth respond slightly to temperature in the research area. Anthropogenic activities also caused changes in the plant cover and NDVI of the study area. In this regard, we can see how the growing rate of mean temperature and decreasing rate of

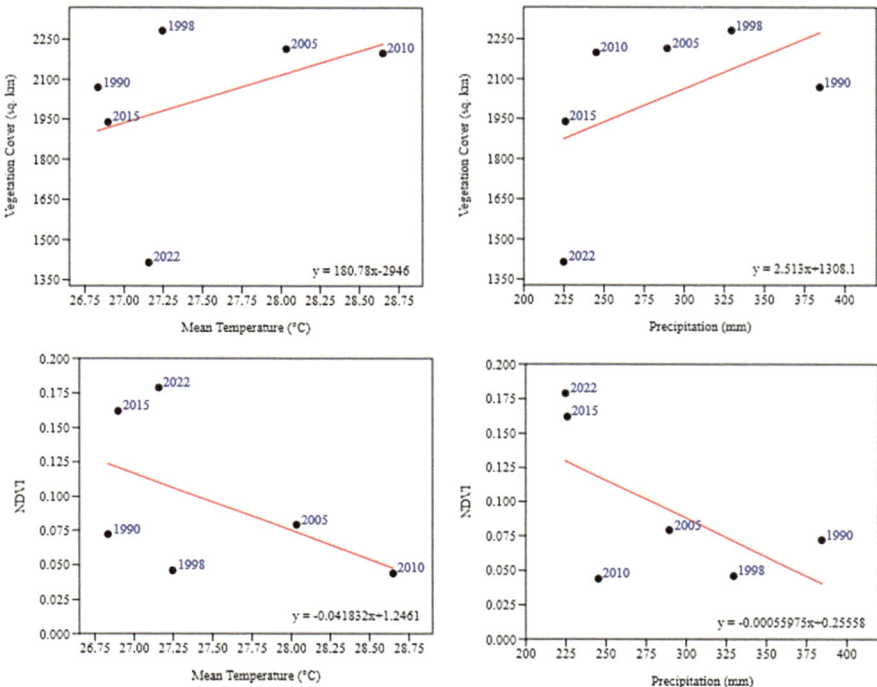

Fig. 9.24 Relationship between climatic variables (temperature and precipitation), NDVI, and vegetation coverage (pre-monsoon season)

mean precipitation, as well as their geographical variability, influenced the mean pixel values of NDVI and the decreasing area of vegetation cover in the study area.

Post-Monsoon Season

Table 9.4 presents the values of post-monsoon seasonal climate data (temperature and precipitation), NDVI (Fig. 9.22), and vegetation cover (Fig. 9.23) from 1990 to 2022. Figure 9.25 displays the link between climatic variables (temperature and precipitation), vegetation features, and vegetation cover. Figure 9.25 shows the graphical relationship between climatic variables, NDVI, and vegetation cover in the study area during the post-monsoon season. The temperature and vegetation NDVI regression research demonstrates a moderate range of regression coefficient values between 0.4834 and 0.2088 for the precipitation-NDVI relationship. Similarly, regression analysis of temperature, precipitation, and vegetation coverage reveals low regression coefficients of 0.0761, 0.0449, and 0.0449, respectively.

Therefore, the monsoonal precipitation sufficiently meets the water requirements of plants during the post-monsoon season. According to present research, precipitation helps to maintain adequate moisture availability in soil, allowing plant roots to efficiently draw water from the soil for growth. In this regard, the demonstrated points for 2015 and 2022 show an intense connection with both climatic factors. In

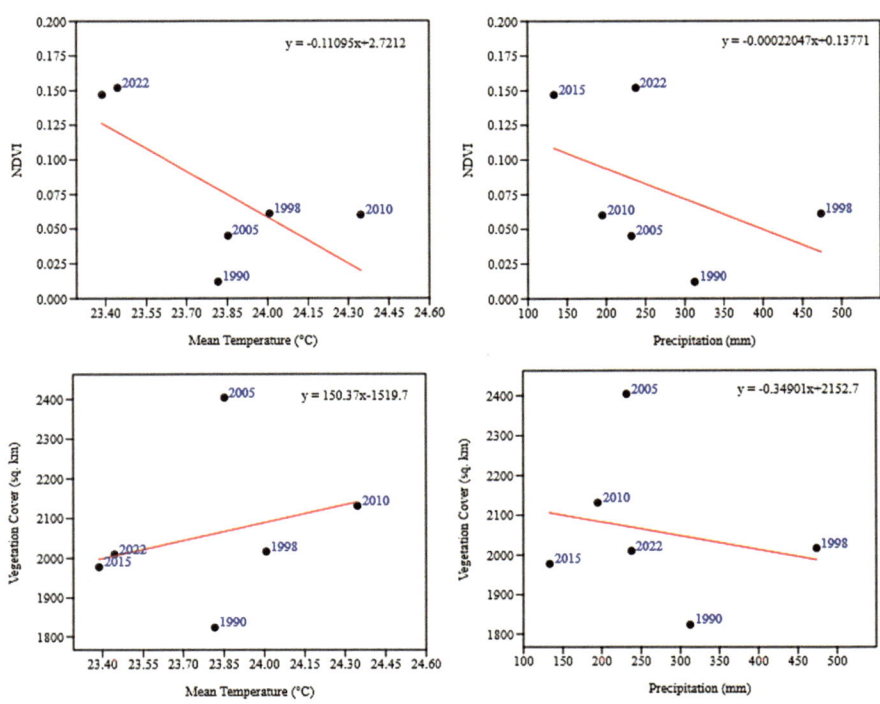

Fig. 9.25 Relationship between climatic variables (temperature and precipitation), NDVI, and vegetation coverage (post-monsoon season)

Table 9.4 Values of NDVI, vegetation cover, and climatic parameters of coastal 24 Parganas (S) (post-monsoon season)

Year	NDVI	Vegetation cover (sq. km)	Mean temperature (°C)	Precipitation (mm)
1990	0.012	1824.230	23.816	312.353
1998	0.061	2016.545	24.007	473.842
2005	0.045	2403.534	23.853	231.838
2010	0.060	2131.458	24.346	195.065
2015	0.147	1977.694	23.388	133.214
2022	0.152	2009.864	23.445	237.767

Source: Calculated by the author

the current study area, temperature reacted inversely to vegetation cover. It appears that rising temperatures reduced soil moisture availability in the post-monsoon season and decreased vegetation water requirements, resulting in low vegetation growth.

From April to June 2022, this study surveyed 300 respondents to assess local perspectives on climate change and its impact on mangrove health. Summarized findings are presented in Table 9.5, while Fig. 9.26, based on Garrett analysis, highlights the most significant factors affecting mangrove health risks. Climate change/variability ranked first, followed by climate disasters, diseases, human activities, lack of governance, and insect attacks. Notably, climate change/variability received the highest attention, with a Garrett average score of 73.57. Specifically, rising temperatures, shifting precipitation patterns, and sea level rise (Kanan et al., 2023) are key drivers of climate-related risks impacting coastal vegetation, including mangrove ecosystems. Additionally, research suggests a direct correlation between climate change/variability and human activities (Ghosh & Roy, 2022; Sen, 2019), such as natural disasters, diseases, and insect invasions.

In summary, we concluded with the following observations: Climate has a somewhat significant impact on vegetation NDVI and coverage in the research area, which may be tracked and predicted with satellite imagery. Furthermore, participatory methodologies can be utilized to better understand local technical knowledge and opinions concerning vegetation health risk assessment and associated actions, which will aid in the development of future research and decision-making. As a result, we may conclude that climatic parameter variability causes the changeability of vegetation in the coastal socio-ecological system, as well as the alteration of these domains.

Table 9.5 Respondent opinions of climate conditions and their impact on vegetation health risks

Concerns about climatic parameters and vegetation health risks	Respondents opined "yes" (of %)
Climate change	98.78
Increased temperature	94.06
Decreased temperature	05.94
Increased precipitation/pattern	0.59
Decreased precipitation	99.41
Sea level rise (+ve)	100.00
Increased cyclonic activity	100.00
Decreased cyclonic activity	00.00
Increased anthropogenic activities in the forest land	100.00
Degradation of mangrove ecosystem (-ve)	97.35
Climate change degraded mangrove ecosystems	88.73
Anthropogenic activities degraded mangrove ecosystems	62.96

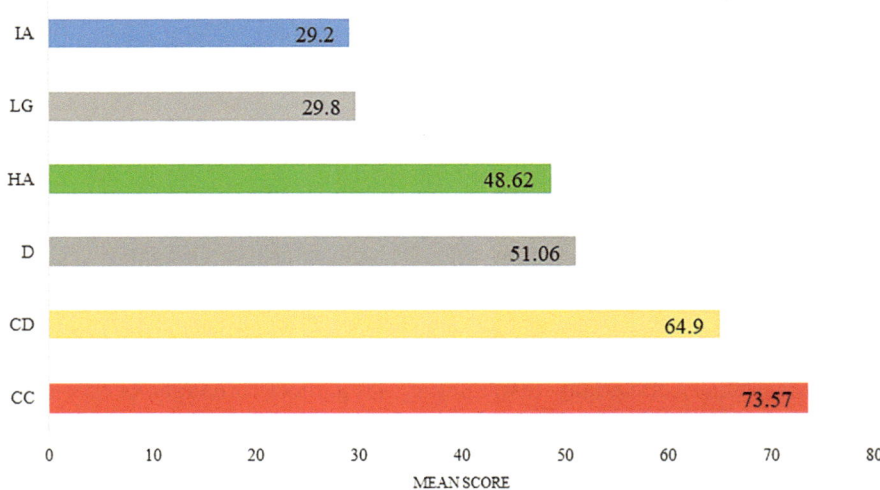

Fig. 9.26 Respondents ranked the distribution of Garrett mean scores. (CC: climate change/variability; CD: climate disasters; D: diseases; HA: human activities; LG: less governance; IA: insect attack)

9.3   Conclusion and Arguments

In conclusion, this chapter mapped local stakeholders' knowledge and employed PGIS tools and methodologies in various forms of analysis, as well as involved individuals in spatially mapping their surroundings, concerns, and resources. Citizens use PGIS to actively participate in the collection, analysis, and visualization of geographic data, bringing vital insights into their communities. This technique promotes geospatial citizenship by enabling individuals to participate actively in local decision-making processes. Moreover, citizens gain a better grasp of their surroundings and may advocate for their needs when characteristics such as infrastructure, environmental risks, social amenities, and cultural landmarks are mapped. It also encourages collaboration among community members, government agencies, nongovernmental organizations (NGOs), and researchers, allowing for debate and collective action to address local concerns. It also improves openness and accountability in governance by democratizing access to spatial data, encouraging participatory decision-making, and demonstrating the power of participatory GIS.

References

Abdel-Raheem, S. T. (2019). *Using participatory mapping for community-based marine spatial planning in St. George's Caye, Belize*. Thesis submitted to Geographic Information Science, San Francisco State University, San Francisco, California. https://scholarworks.calstate.edu/downloads/f4752j56n

Acharya, G. (2002). Life at the margins: The social, economic and ecological importance of mangroves. *Madera y Bosques, 8*(1), 53–60. https://www.redalyc.org/pdf/617/61709803.pdf

Akram, H., Hussain, S., Mazumdar, P., Chua, K. O., Butt, T. E., & Harikrishna, J. A. (2023). Mangrove health: A review of functions, threats, and challenges associated with mangrove management practices. *Forests, 14*(9), 1698. https://doi.org/10.3390/f14091698

Andrade-Sánchez, J., Eaton-Gonzalez, R., Leyva-Aguilera, C., & Wilken-Robertson, M. (2021). Indigenous mapping for integrating traditional knowledge to enhance community-based vegetation management and conservation: The Kumeyaay basket weavers of San José de la Zorra, México. *ISPRS International Journal of Geo-Information, 10*(3), 124. https://doi.org/10.3390/ijgi10030124

André, L. V., Van Wynsberge, S., Chinain, M., & Andréfouët, S. (2022). Benefits of collaboration between indigenous fishery management and data-driven spatial planning approaches: The case of a Polynesian traditional design (rāhui). *Fisheries Research, 256*, 106475. https://doi.org/10.1016/j.fishres.2022.106475

Baldwin, K., & Mahon, R. (2014). A participatory GIS for marine spatial planning in the Grenadine Islands. *The Electronic Journal of Information Systems in Developing Countries, 63*(1), 1–18. https://doi.org/10.1002/j.1681-4835.2014.tb00452.x

Barange, M., Bahri, T., Beveridge, M. C. M., et al. (2018). *Impacts of climate change on fisheries and aquaculture: Synthesis of current knowledge, adaptation and mitigation options.* Food and Agriculture Organization of the United Nations. ISBN: 978-92-5-130607-9. https://www.fao.org/3/i9705en/i9705en.pdf

Bhendekar, S. N., Shenoy, L., Raje, S. G., Chellappan, A., & Singh, R. (2016). Participatory GIS in trawl fisheries along Mumbai coast, Maharashtra. *Indian Journal of Geo-Marine Sciences, 45*(8), 937–942.

Boeser, S. M., & Hamylton, S. M. (2019). Geographic Information Systems (GIS). In C. W. Finkl & C. Makowski (Eds.), *Encyclopedia of coastal science* (Encyclopedia of earth sciences series). Springer. https://doi.org/10.1007/978-3-319-93806-6_149

Brown, G., & Kyttä, M. (2014). Key issues and research priorities for public participation GIS (PPGIS): A synthesis based on empirical research. *Applied Geography, 46*, 122–136. https://doi.org/10.1016/j.apgeog.2013.11.004

Carmen, E., Fazey, I., Ross, H., et al. (2022). Building community resilience in a context of climate change: The role of social capital. *Ambio, 51*, 1371–1387. https://doi.org/10.1007/s13280-021-01678-9

Carugati, L., Gatto, B., Rastelli, E., et al. (2018). Impact of mangrove forests degradation on biodiversity and ecosystem functioning. *Scientific Reports, 8*, 13298. https://doi.org/10.1038/s41598-018-31683-0

da Silva, C. V., Ortigao, M., Willaert, T., Rosa, R., Nunes, L. C., & Cunha-e-Sa, M. A. (2021). Participatory Geographic Information Systems (PGIS): Alternative approaches to identify potential conflicts and positional accuracy in marine and coastal ecosystem services. *Marine Policy, 131*, 104650. https://doi.org/10.1016/j.marpol.2021.104650

Dasgupta, D. (2018). Homes, lives are swept away. *The Straits Times.* https://www.straitstimes.com/asia/south-asia/homes-lives-are-swept-away.

Dawit, D., & Simane, B. (2017). Community forest management for climate change mitigation and adaptation in Ethiopia: Determinants of community participation. In W. Leal Filho, S. Belay, J. Kalangu, W. Menas, P. Munishi, & K. Musiyiwa (Eds.), *Climate change adaptation in Africa. Climate change management.* Springer. https://doi.org/10.1007/978-3-319-49520-0_27

Dineshbabu, A. P., Thomas, S., Rohit, P., & Maheswarudu, G. (2019). Marine spatial planning for resource conservation, fisheries management and for ensuring fishermen security–global perspectives and Indian initiatives. *Current Science, 116*(4), 561–567.

Diswandi, D. (2022). Community forestry Management for Climate Change Adaptation. In M. Lackner, B. Sajjadi, & W. Y. Chen (Eds.), *Handbook of climate change mitigation and adaptation.* Springer. https://doi.org/10.1007/978-3-030-72579-2_135

Edwin, L., Sankar, T. V., & Pillai, D. (Eds.). (2019). *Traditional knowledge and management systems in fisheries.* Society of Fisheries Technologists. ISBN: 978-81-934622-0-1.

Eilola, S., Fagerholm, N., Mäki, S., Khamis, M., & Käyhkö, N. (2014). Realization of participation and spatiality in participatory forest management—A policy–practice analysis from Zanzibar, Tanzania. *Journal of Environmental Planning and Management, 58*, 1242–1269. https://doi. org/10.1080/09640568.2014.921142

Escandón-Panchana, J., Elao Vallejo, R., Escandón-Panchana, P., Velastegui-Montoya, A., & Herrera-Franco, G. (2022). Spatial planning of the coastal marine socioecological system—Case study: Punta Carnero, Ecuador. *Resources, 11*(8), 74. https://doi.org/10.3390/ resources11080074

Fagerholm, N., Eilola, S., Kisanga, D., et al. (2019). Place-based landscape services and potential of participatory spatial planning in multifunctional rural landscapes in southern highlands, Tanzania. *Landscape Ecology, 34*, 1769–1787. https://doi.org/10.1007/s10980-019-00847-2

Fagerholm, N., Raymond, C. M., Olafsson, A. S., Brown, G., Rinne, T., Hasanzadeh, K., et al. (2021). A methodological framework for analysis of participatory mapping data in research, planning, and management. *International Journal of Geographical Information Science, 35*(9), 1848–1875. https://doi.org/10.1080/13658816.2020.1869747

Gabriele, M., Brumana, R., Previtali, M., et al. (2023). A combined GIS and remote sensing approach for monitoring climate change-related land degradation to support landscape preservation and planning tools: The Basilicata case study. *Applied Geomatics, 15*, 497–532. https:// doi.org/10.1007/s12518-022-00437-z

Gandhi, G. M., Parthiban, B. S., Thummalu, N., & Christy, A. (2015). Ndvi: Vegetation change detection using remote sensing and gis–A case study of Vellore District. *Procedia Computer Science, 57*, 1199–1210. https://doi.org/10.1016/j.procs.2015.07.415

Garrett, H. E. (1979). *Statistics in psychology and education* (p. 329). Vakils, Feffer and Simons Ltd.. (9th Indian Reprnt).

Ghosh, S., & Mistri, B. (2023). Cyclone-induced coastal vulnerability, livelihood challenges and mitigation measures of Matla–Bidya inter-estuarine area, Indian Sundarban. *Natural Hazards, 116*, 3857–3878. https://doi.org/10.1007/s11069-023-05840-2

Ghosh, S., & Roy, S. (2022). Climate change, ecological stress and livelihood choices in Indian Sundarban. In A. K. E. Haque, P. Mukhopadhyay, M. Nepal, & M. R. Shammin (Eds.), *Climate change and community resilience*. Springer. https://doi.org/10.1007/978-981-16-0680-9_26

Goodchild, M. F., & Haining, R. P. (2004). GIS and spatial data analysis: Converging perspectives. In R. J. G. M. Florax & D. A. Plane (Eds.), *Fifty years of regional science. Advances in spatial science*. Springer. https://doi.org/10.1007/978-3-662-07223-3_16

Gopal, B., & Chauhan, M. (2006). Biodiversity and its conservation in the Sundarban Mangrove ecosystem. *Aquatic Sciences, 68*(3), 338–354. https://doi.org/10.1007/s00027-006-0868-8

Grati, F., Azzurro, E., Scanu, M., Tassetti, A. N., Bolognini, L., Guicciardi, S., et al. (2022). Mapping small-scale fisheries through a coordinated participatory strategy. *Fish and Fisheries, 23*(4), 773–785. https://doi.org/10.1111/faf.12644

Hemani, C. (2014). Approaches to climate change adaptation of vulnerable coastal communities of India. In W. Leal Filho (Ed.), *Handbook of climate change adaptation*. Springer. https://doi. org/10.1007/978-3-642-40455-9_100-1

Islam, M. M., Sallu, S., Hubacek, K., et al. (2014). Vulnerability of fishery-based livelihoods to the impacts of climate variability and change: Insights from coastal Bangladesh. *Regional Environmental Change, 14*, 281–294. https://doi.org/10.1007/s10113-013-0487-6

Jana, A., Maiti, S., & Biswas, A. (2016). Seasonal change monitoring and mapping of coastal vegetation types along Midnapur-Balasore Coast, Bay of Bengal using multi-temporal landsat data. *Modeling Earth Systems and Environment, 2*, 7. https://doi.org/10.1007/s40808-015-0062-x

Janßen, H., Bastardie, F., Eero, M., Hamon, K. G., Hinrichsen, H. H., Marchal, P., et al. (2018). Integration of fisheries into marine spatial planning: Quo vadis? *Estuarine, Coastal and Shelf Science, 201*, 105–113. https://doi.org/10.1016/j.ecss.2017.01.003

Jayawardhan, S. (2017). Vulnerability and climate change induced human displacement. *Consilience, 17*(1), 103–142.

Johnson, D., Blackett, P., Allison, A. E., & Broadbent, A. M. (2023). Measuring social vulnerability to climate change at the coast: Embracing complexity and context for more accurate and equitable analysis. *Water, 15*(19), 3408. https://doi.org/10.3390/w15193408

Kanan, A. H., Pirotti, F., Masiero, M., et al. (2023). Mapping inundation from sea level rise and its interaction with land cover in the Sundarbans mangrove forest. *Climatic Change, 176*, 104. https://doi.org/10.1007/s10584-023-03574-5

Keenan, R. J. (2015). Climate change impacts and adaptation in forest management: A review. *Annals of Forest Science, 72*, 145–167. https://doi.org/10.1007/s13595-014-0446-5

Kirilenko, A. P. (2022). Geographic Information System (GIS). In R. Egger (Ed.), *Applied data science in tourism. Tourism on the verge*. Springer. https://doi.org/10.1007/978-3-030-88389-8_24

Kitolelei, S., Lowry, J. H., Qaqara, N., Ryle, J., Veitayaki, J., & Piovano, S. (2022). Spatial use of marine resources in a rural village: A case study from Qoma, Fiji. *Frontiers in Marine Science, 9*, 993103. https://doi.org/10.3389/fmars.2022.993103

Kwaku Kyem, P. A. (2004). Of intractable conflicts and participatory GIS applications: The search for consensus amidst competing claims and institutional demands. *Annals of the Association of American Geographers, 94*(1), 37–57. https://doi.org/10.1111/j.1467-8306.2004.09401003.x

Lamptey, F. (2009). *Participatory GIS tools for mapping indigenous knowledge in customary land tenure dynamics case of Peri-urban northern Ghana*. Thesis submitted to the International Institute for Geo-information Science and Earth Observation Enschede. https://webapps.itc.utwente.nl/librarywww/papers_2009/msc/gimla/lamptey.pdf

Levine, A. S., & Feinholz, C. L. (2015). Participatory GIS to inform coral reef ecosystem management: Mapping human coastal and ocean uses in Hawaii. *Applied Geography, 59*, 60–69. https://doi.org/10.1016/j.apgeog.2014.12.004

Lv, Y., Sarker, M. N. I., & Firdaus, R. R. (2024). Disaster resilience in climate-vulnerable community context: Conceptual analysis. *Ecological Indicators, 158*, 111527. https://doi.org/10.1016/j.ecolind.2023.111527

Macnab, P. (2002). There must be a catch: Participatory GIS in a Newfoundland fishing community. Chapter 13. In *Community participation and geographical information systems* (pp. 173–191). CRC Press. https://doi.org/10.1201/9780203469484. ISBN: 9780429203961.

Madhav, S., Nazneen, S., & Singh, P. (Eds.). (2022). *Coastal ecosystems: Environmental importance, current challenges and conservation measures*. Springer. https://doi.org/10.1007/978-3-030-84255-0. ISBN: 978-3-030-84255-0.

Malakar, K. D. (2020). Rural livelihood and Mangrove degradation: A case study of Namkhana Block, West Bengal, India. *International Journal of Innovative Science and Research Technology, 5*(1), 721–726.

Massey, R., Berner, L. T., Foster, A. C., Goetz, S. J., & Vepakomma, U. (2023). Remote sensing tools for monitoring forests and tracking their dynamics. In M. M. Girona, H. Morin, S. Gauthier, & Y. Bergeron (Eds.), *Boreal forests in the face of climate change. Advances in global change research* (Vol. 74). Springer. https://doi.org/10.1007/978-3-031-15988-6_26

McCall, M. K., & Dunn, C. E. (2012). Geo-information tools for participatory spatial planning: Fulfilling the criteria for "good" governance? *Geoforum, 43*, 81–94. https://doi.org/10.1016/j.geoforum.2011.07.007

McCall, M. K., & Minang, P. A. (2005). Assessing participatory GIS for community-based natural resource management: Claiming community forests in Cameroon. *Geographical Journal, 171*(4), 340–356.

Mitra, A. (2013). *Sensitivity of mangrove ecosystem to changing climate*. Springer. https://doi.org/10.1007/978-81-322-1509-7. ISBN: 978-81-322-1509-7.

Muhala, V., Chicombo, T. F., Macate, I. E., Guimarães-Costa, A., Gundana, H., Malichocho, C., et al. (2021). Climate change in fisheries and aquaculture: Analysis of the impact caused by Idai and Kenneth cyclones in Mozambique. *Frontiers in Sustainable Food Systems, 5*, 714187. https://doi.org/10.3389/fsufs.2021.714187

Nunes, A. R. (2021). Exploring the interactions between vulnerability, resilience and adaptation to extreme temperatures. *Natural Hazards, 109*, 2261–2293. https://doi.org/10.1007/s11069-021-04919-y

Nutters, H. (2012). *Addressing social vulnerability and equity in climate change adaptation planning*. San Francisco Bay Conservation and Development Commission. Available at: https://www.adaptingtorisingtides.org/wp-content/uploads/2015/04/ART_Equity_WhitePaper.pdf

Pandya, U., Mudaliar, A. N., & Alvi, S. (2023). Coastal vegetation change detection using a remote sensing approach. *Environmental Sciences Proceedings, 29*(1), 17. https://doi.org/10.3390/ECRS2023-15853

Pattanaik, A. (2021). *Mangrove-dependent small-scale fisher (SSF) communities in the Sundarbans—Vulnerable yet viable*. Thesis submitted to University of Waterloo, Ontario, Canada. https://uwspace.uwaterloo.ca/bitstream/handle/10012/17366/Pattanaik_Aishwarya.pdf?sequence=3

Paul, S., Mishra, M., Pati, S., Acharyya, T., Santos, C. A. G., da Silva, R. M., et al. (2024). Evaluation of overwash vulnerability and shoreline dynamics in cyclone-prone Sagar Island, Sundarbans (India). *Science of the Total Environment, 907*, 167933. https://doi.org/10.1016/j.scitotenv.2023.167933

Paulangan, Y. P., Rumbiak, K., & Barapadang, B. (2020). Fishing season and participatory mapping of the fishing ground of target fish in Depapre Bay, Jayapura regency, Papua Indonesia. In *IOP Conference Series: Earth and Environmental Science* (Vol. 584, No. 1, p. 012031). IOP Publishing. https://doi.org/10.1088/1755-1315/584/1/012031

Ponnusamy, K., Gupta, J., & Nagarajan, R. (2009). Indigenous Technical Knowledge (ITKs) in dairy enterprise in coastal Tamil Nadu. *Indian Journal of Traditional Knowledge, 8*(2), 206–211.

Poto, M. P., Kuhn, A., Tsiouvalas, A., et al. (2022). Knowledge integration and good marine governance: A multidisciplinary analysis and critical synopsis. *Human Ecology, 50*, 125–139. https://doi.org/10.1007/s10745-021-00289-y

Priyadarshini, S. (2015). Climate change pushing Sundarban farmers into 'awkward jobs'. *Nature India*. https://doi.org/10.1038/nindia.2015.21

Rakotomahazo, C., Ravaoarinorotsihoarana, L. A., Randrianandrasaziky, D., Glass, L., Gough, C., Todinanahary, G. G. B., & Gardner, C. J. (2019). Participatory planning of a community-based payments for ecosystem services initiative in Madagascar's mangroves. *Ocean & Coastal Management, 175*, 43–52. https://doi.org/10.1016/j.ocecoaman.2019.03.014

Ramsey, K. (2008). A call for agonism: GIS and the politics of collaboration. *Environment and Planning A, 40*(10), 2346–2363. https://doi.org/10.1068/a4028

Reddy, G. P. O. (2018). Spatial data management, analysis, and modeling in GIS: Principles and applications. In G. Reddy & S. Singh (Eds.), *Geospatial technologies in land resources mapping, monitoring and management. Geotechnologies and the environment* (Vol. 21). Springer. https://doi.org/10.1007/978-3-319-78711-4_7

Roy, A., Sinha, A., Manna, R. K., Aftabuddin, M. D., & Das, S. K. (2020). Traditional knowledge of the fishermen community of Indian Sundarbans: An assessment of rationality and effectiveness. *Indian Journal Of Fisheries, 67*(2), 94–101. https://doi.org/10.21077/ijf.2019.67.2.86752-13

Said, A., & Trouillet, B. (2020). Bringing 'deep knowledge' of fisheries into marine spatial planning. *Maritime Studies, 19*, 347–357. https://doi.org/10.1007/s40152-020-00178-y

Salvatteci, R., Schneider, R. R., Field, E. G. D., et al. (2022). Smaller fish species in a warm and oxygen-poor Humboldt current system. *Science, 375*, 101–104. https://doi.org/10.1126/science.abj0270

Sandilyan, S., & Kathiresan, K. (2012). Mangrove conservation: A global perspective. *Biodiversity and Conservation, 21*, 3523–3542. https://doi.org/10.1007/s10531-012-0388-x

Sen, H. S. (Ed.). (2019). *The Sundarbans: A disaster-prone eco-region: Increasing livelihood security*. Coastal Research Library, Springer International Publishing. https://doi.org/10.1007/978-3-030-00680-8. ISBN: 978-3-030-00680-8.

Shenoy, A. S. (2009). Indigenous technical knowledge and its relevance for sustainability. *Resource Papers, 87th Foundation Course for Agricultural Research Service* (pp. 541–549). ICAR-National Academy for Agriculture Research Management, .

Shiiba, N., Singh, P., Charan, D., et al. (2023). Climate change and coastal resiliency of Suva, Fiji: A holistic approach for measuring climate risk using the climate and ocean risk vulnerability index (CORVI). *Mitig Adapt Strateg Glob Change, 28*, 9. https://doi.org/10.1007/s11027-022-10043-4

Shivanna, K. R. (2022). Climate change and its impact on biodiversity and human welfare. *Proceedings of the National Academy of Sciences, 88*, 160–171. https://doi.org/10.1007/s43538-022-00073-6

Shyam, S. S., & Antony, P. (2013). Indigenous Technical Knowledge (ITK) in capture fisheries: A case study in Vypeen Island of Ernakulam district. *Discovery Nature, 4*(11), 7–10.

Sieber, R. (2006). Public participation geographic information systems: A literature review and framework. *Annals of the Association of American Geographers, 96*(3), 491–507. https://doi.org/10.1111/j.1467-8306.2006.00702.x

Stelzenmüller, V., Gimpel, A., Gopnik, M., & Gee, K. (2017). Aquaculture site-selection and marine spatial planning: The roles of GIS-based tools and models. In B. Buck & R. Langan (Eds.), *Aquaculture perspective of multi-use sites in the Open Ocean*. Springer. https://doi.org/10.1007/978-3-319-51159-7_6

Sullivan-Wiley, K. A., Gianotti, A. G. S., & Connors, J. P. C. (2019). Mapping vulnerability: Opportunities and limitations of participatory community mapping. *Applied Geography, 105*, 47–57. https://doi.org/10.1016/j.apgeog.2019.02.008

Termorshuizen, J. W., & Opdam, P. (2009). Landscape services as a bridge between landscape ecology and sustainable development. *Landscape Ecology, 24*, 1037–1052. https://doi.org/10.1007/s10980-008-9314-8

Tripathi, N., & Bhattarya, S. (2004). Integrating indigenous knowledge and GIS for participatory natural resource management: State-of-the-practice. *EJISDC, 17*(3), 1–13. https://doi.org/10.1002/j.1681-4835.2004.tb00112.x

White, B. P., Breakey, S., Brown, M. J., Smith, J. R., Tarbet, A., Nicholas, P. K., & Ros, A. M. V. (2023). Mental health impacts of climate change among vulnerable populations globally: An integrative review. *Annals of Global Health, 89*(1). https://doi.org/10.5334/aogh.4105

World Bank. (2014). *Building resilience for sustainable development of the Sundarbans: Strategy report (No. 20116; World Bank Other Operational Studies)*. The World Bank. https://ideas.repec.org/p/wbk/wboper/20116.html

Yamamoto, Y. (2023). Living under ecosystem degradation: Evidence from the mangrove–fishery linkage in Indonesia. *Journal of Environmental Economics and Management, 118*, 102788. https://doi.org/10.1016/j.jeem.2023.102788

Part III
The Future of Participatory GIS in Social Sciences Research

Chapter 10
The Power of Participatory GIS: Disciplinary Principles and Research Scope

Where discipline and conversation intersect, a collaborative
space emerges for communities to actively participate in the
co-creation of their geographic realities.

K.D. Malakar (author)

Abstract In this chapter, we will take an exploration into the dynamic world of participatory GIS (PGIS). This chapter aims to unravel the dense tapestry of disciplinary principles that support PGIS, providing readers with a complete understanding of its conceptual foundations, analytical complexities, and future research opportunities in the social sciences branches. This cutting-edge strategy, through its integration with PGIS, empowers communities and stakeholders to actively participate in the development, interpretation, and application of geographical information. Here, we highlight the power of PGIS in the field of social sciences. It is also a collective call to action, encouraging academics, practitioners, and policymakers to adopt a more democratic and inclusive approach to spatial knowledge generation. Through these talks, we suggest readers immerse themselves in the rich tapestry of participatory GIS, where the convergence of technology and community participation opens up new possibilities for our knowledge of space, place, and the power dynamics that define our society.

Keywords Disciplinary nature of participatory GIS · Social sciences · Societal space · Future of PGIS · Social sciences research and PGIS

In this chapter, we delve into the world of participatory GIS (PGIS), aiming to uncover the foundational principles, complexities, and future opportunities in the social sciences. This cutting-edge approach empowers communities to actively contribute to geographical information, emphasizing its significance in the social sciences. We call for collective action, urging academics, practitioners, and policymakers to embrace a more democratic and inclusive spatial knowledge generation approach. Through these discussions, we invite readers to explore the synergy of technology and community participation in participatory GIS, offering new possibilities for understanding space, place, and societal power dynamics.

Key Points of the Chapter
- Understanding the disciplinary principles and diverse fields of study of participatory GIS in the social sciences around the world.
- The effectiveness of participatory GIS in social science research.

10.1   Introduction

Participatory GIS functions at the intersection of technology and social science, providing a strong lens through which to analyze, comprehend, and address complex socio-spatial challenges. PGIS, focused on the ideas of inclusivity, collaboration, and community involvement, goes beyond typical GIS approaches by putting the power of spatial information directly in the hands of individuals who live, work, and interact in a specific region. At its foundation, the social science academic concepts of PGIS are based on the idea that spatial information is not neutral but has social, cultural, and political implications.

PGIS research covers a wide range of social scientific disciplines, including anthropology, sociology, geography, urban studies, and environmental science. This multidisciplinary nature reflects the recognition that socio-spatial processes are inherently complex and multidimensional, needing a comprehensive approach that goes beyond traditional disciplinary boundaries. As an outcome, the following discussion (Sect. 10.2) addresses the many disciplinary principles and study scope of PGIS in the social sciences.

10.2   Disciplinary Principles and Research Scope

Table 10.1 encompasses several disciplinary ideas and research areas of PGIS in the social sciences.

10.3   Conclusion and Arguments

In conclusion, participatory GIS is a strong and transformational instrument in the field of social sciences. Its application in a variety of fields, including sociology, history, geography, political science, socio-ecological studies, and public administration, shows a paradigm shift toward inclusive and community-driven research approaches. The principles of participatory GIS, which are based on community engagement, ethical data processing, and collaborative decision-making,

Table 10.1 Disciplinary principles and research scope of participatory GIS

Sr. no.	PGIS and social sciences disciplines	Disciplinary principles	Research scope
10.2.1	PGIS in anthropology	*Cultural context:* focuses on recognizing local cultural variations in PGIS operations *Ethnographic sensitivity:* uses ethnographic approaches to gain a comprehensive sociocultural understanding *Community collaboration:* emphasizes the active participation of communities in data collection, analysis, and decision-making *Knowledge co-creation:* promotes collaborative knowledge development among scholars and communities	*Cultural mapping:* uses PGIS to map cultural landscapes and heritage *Indigenous knowledge preservation:* helps to preserve and document indigenous knowledge via spatial representation *Socio-spatial analysis:* it helps to investigate social dynamics using spatial patterns and linkages *Community empowerment:* using PGIS to empower communities in decision-making
10.2.2	PGIS in archaeology	*Site documentation:* archaeological sites are analyzed and documented using PGIS *Data visualization:* makes use of PGIS to visualize and analyze archaeological data *Contextual understanding:* uses PGIS in conjunction with historical settings to comprehend spatial linkages *Collaborative:* collaborative research places a strong emphasis on working together with stakeholders and local communities	*Strategic planning:* utilizing PGIS for strategic planning during archaeological excavations is known as excavation planning *Local participation:* engages local people in the mapping and interpretation of the site *Investigation of spatial patterns:* analyzing spatial patterns can improve our understanding of archaeology *Heritage sites:* cultural heritage sites are managed and preserved through the use of PGIS in heritage management
10.2.3	PGIS in area studies	*Contextual understanding:* highlights how crucial it is for spatial analysts to comprehend local circumstances *Multidisciplinary approach:* integrates knowledge from multiple fields within area studies to adopt a comprehensive viewpoint *Cultural sensitivity:* ensures sensitivity in PGIS applications within designated areas by acknowledging and respecting cultural diversity *Community integration:* including the local community in the mapping and interpreting process is known as community involvement	*Mapping cultural landscapes:* examines the use of PGIS in the mapping and preservation of cultural landscapes within certain regions *Regional planning:* uses PGIS to support well-informed regional planning and development *Conflict resolution:* examines and resolves social, economic, or environmental disputes in the areas under study by using spatial data *Community-based resource management:* uses PGIS technology to manage resources sustainably by involving communities

(continued)

Table 10.1 (continued)

Sr. no.	PGIS and social sciences disciplines	Disciplinary principles	Research scope
10.2.4	PGIS in communication studies	*Audience engagement:* GIS is used in conjunction with audience engagement to comprehend the spatial patterns of media consumption and audience involvement *Media mapping:* maps the distribution and content of media using GIS in order to do communication research *Community input:* uses GIS to involve communities in a collaborative review of communication tactics *Visual communication analysis:* spatial data integration is used in visual communication analysis to examine the visual components of communication materials	*Media accessibility:* examines GIS uses to improve media content accessibility *Digital inclusion:* GIS is looked at in the context of digital inclusion in order to identify digital disparities and promote inclusive communication infrastructures *Spatial representation of narratives:* investigates the use of GIS for the analysis and spatial representation of communication narratives *Community media planning:* using GIS cooperation, communities are involved in the planning of media activities
10.2.5	PGIS in consumer studies	*Retail site selection:* uses PGIS to help retail enterprises choose key locations *Market analysis:* uses PGIS to examine customer behavior and preferences *Demographic mapping:* uses PGIS to incorporate demographic data for focused customer insights *Customer involvement:* involves customers in PGIS procedures to improve data relevance and accuracy	*Market accessibility:* it looks into using PGIS to map marketplaces and enhance customer access to services *Shopping behavior analysis:* examines the use of PGIS to examine spatial trends in consumer purchasing behavior *Customer experience mapping:* maps and comprehends the spatial components of the consumer experience using PGIS *Consumer feedback mapping:* customer feedback mapping involves customers in a spatial mapping process to improve the development of products and services
10.2.6	PGIS in criminology	*Environmental criminology:* environmental criminology uses geographical analysis to comprehend how the surroundings affect criminal behavior *Crime mapping:* maps and analyzes spatial patterns of crime episodes using PGIS *Community policing:* community policing involves involving communities in PGIS procedures in order to work together to address criminal challenges *Social justice:* uses PGIS to map crime-affected areas, placing a strong emphasis on social justice theory in criminological research	*Community safety planning:* makes use of PGIS to facilitate cooperative community involvement in the creation of crime prevention plans *Crime hotspot identification:* investigates PGIS applications for the purpose of identifying and geographically analyzing crime hotspots *Programs for the reentry of prisoners:* examines the spatial dimensions of resource mapping for the successful reintegration of prisoners into society *Policing resource allocation:* using spatial analysis, PGIS is examined to maximize resource allocation in law enforcement

10.2.7	PGIS in cultural studies	Cultural mapping: PGIS is used to map cultural manifestations and phenomena spatially Critical cultural analysis: using PGIS, critical cultural analysis examines the geographical aspects of cultural dynamics and identity Community engagement: promotes community participation in PGIS procedures by appreciating local viewpoints Heritage preservation: cultural heritage locations and traditions are documented and preserved through the integration of PGIS in heritage preservation	Cultural identification mapping: it is a PGIS-based method for representing and comprehending the spatial aspects of cultural identification Cultural landscape analysis: examines how to map and analyze cultural landscapes using PGIS technologies Community-based heritage management: communities are involved in PGIS for the purpose of managing and conserving cultural resources through community-based heritage management Cultural impact assessment: examines the use of PGIS to evaluate the spatial effects of cultural practices on communities and environments
10.2.8	PGIS in Demography	Community involvement: promotes communities' active involvement in the methods used to gather demographic data Population mapping: makes use of PGIS to map and analyze population data Monitoring of demographic change: uses PGIS to track changes in a region's population in real time Spatial distribution analysis: analyzing and visualizing spatial trends in population distribution with the integration of PGIS is known as spatial distribution analysis	Impact of urbanization: examines PGIS to comprehend how urbanization affects demographic trends spatially Healthcare access: assesses and enhances spatial access to healthcare services through the use of PGIS Migration trends: examines how to map and examine geographical trends in population migration using PGIS technologies Community-based planning: uses PGIS to involve local communities in cooperative resource allocation and demographic planning
10.2.9	PGIS in development studies	Holistic development analysis: uses PGIS to conduct a comprehensive analysis of economic, social, and environmental components of development Community-centered development: promotes long-term development by encouraging community participation in PGIS procedures Local empowerment: uses PGIS to empower communities by embracing their knowledge and needs Inclusive decision-making: incorporates various stakeholders into PGIS applications to promote inclusive decision-making	Natural resource management: investigates the use of PGIS for sustainable development via participatory natural resource mapping Poverty mapping: investigates PGIS applications for mapping and analyzing geographical aspects of poverty Infrastructure planning: uses PGIS to plan and optimize infrastructure development in partnership with communities Community-led projects: communities use PGIS to create and implement development projects that address local needs

(continued)

Table 10.1 (continued)

Sr. no.	PGIS and social sciences disciplines	Disciplinary principles	Research scope
10.2.10	PGIS in disability studies	*Inclusive design:* incorporates PGIS into the design process to create accessible environments for people with disabilities *Accessibility mapping:* uses PGIS to map and analyze spatial accessibility for people with disabilities *Barrier identification:* uses PGIS to identify and address spatial barriers that impact people with disabilities *Planning:* participatory planning involves individuals with disabilities in PGIS procedures that inform urban and public space development	*Access mapping:* examines PGIS applications for mapping and assessing the accessibility of public settings for people with impairments *Social inclusion analysis:* uses PGIS to analyze spatial elements that influence social inclusion for people with disabilities *Inclusive transportation planning:* uses PGIS to plan accessible transportation routes and infrastructure *Disability-informed urban design:* involves communities in PGIS to create urban places that meet the requirements of people with impairments
10.2.11	PGIS in economics	*Market analysis:* uses PGIS to analyze market trends and economic activity *Spatial econometrics:* geospatial analysis is used in econometric modeling and forecasting through the use of PGIS in spatial econometrics *Community engagement:* promotes community involvement in the PGIS-based procedures used to gather economic data *Resource allocation:* uses PGIS integration to maximize the distribution of resources in economic planning	*Mapping economic inequality:* maps and analyzes spatial patterns of economic inequality using PGIS *Market accessibility:* examines how to improve and evaluate spatial accessibility to markets through the use of PGIS tools *Economic impact assessment:* uses PGIS to evaluate the spatial effects of development initiatives and economic policies *Community-based business planning:* PGIS is investigated for cooperative community involvement in local business planning as part of community-based business planning
10.2.12	PGIS in education	*Spatial learning:* enhances geography and spatial awareness in education by integrating PGIS for spatial analysis *Community involvement:* promotes students' and communities' active involvement in PGIS initiatives for educational goals *Experiential learning:* uses PGIS to facilitate practical, hands-on learning that develops problem-solving and critical thinking abilities *Accessible curriculum design:* utilizing PGIS, accessible curriculum designers create programs that integrate technology and spatial thinking	*Environmental education:* examines PGIS as a tool for teaching and comprehending environmental issues via spatial analysis *Geographic understanding and spatial literacy:* examines PGIS applications to improve geographic knowledge and spatial literacy *Digital learning aids:* utilizing PGIS, students create and utilize digital maps as interactive and captivating learning aids *Community mapping projects:* as a means of promoting education, PGIS is used in cooperative community mapping projects

10.2.13	PGIS in environmental studies	*Mapping ecosystems*: maps and analyzes geographical patterns in ecosystems using PGIS *Community engagement*: promotes community involvement in PGIS-based environmental data collection procedures *Conservation planning*: utilizing PGIS, conservation methods based on geographic analysis are planned and carried out *Sustainable development*: examines and promotes sustainable development strategies using PGIS, emphasizing their effects on the environment	*Climate change impact assessment*: examines the use of PGIS to evaluate the spatial effects of climate change on ecosystems *Natural resource management*: makes use of PGIS to facilitate cooperative community participation in planning for sustainable natural resources *Biodiversity mapping*: examines how to map and track spatial patterns in biodiversity using PGIS technologies *Environmental justice*: maps and addresses spatial imbalances in access to and quality of the environment by including communities through PGIS
10.2.14	PGIS in ethnic studies	*Cultural mapping*: maps ethnic identity's spatial aspects and cultural landscapes using PGIS *Critical ethnic studies*: PGIS is used in critical ethnic studies to study the geographical aspects of privilege, power, and social justice *Community collaboration*: promotes ethnic communities' active involvement in PGIS data collection and interpretation procedures *Cultural heritage preservation*: uses PGIS to record and conserve the spatial components of ethnic cultural heritage	*Community empowerment*: makes use of PGIS to facilitate cooperative community involvement in the representation and conservation of ethnic history *Identification mapping*: examines how to map and examine the spatial aspects of ethnic identification using PGIS technologies *Cultural impact assessment*: uses PGIS to involve communities in evaluating the spatial effects of cultural norms and policies on ethnic areas *Partial inequality analysis*: examines PGIS to analyze spatial inequities and inequalities that impact ethnic populations
10.2.15	PGIS in family studies	*Family dynamics mapping*: maps and analyzes geographical aspects of family relationships using PGIS *Spatial demographic analysis*: PGIS is integrated into spatial demographic analysis to comprehend spatial trends in family structures and demographics *Community involvement*: creates opportunities for families to actively participate in the procedures of the PGIS data collection and processing *Family support planning*: based on accessibility and spatial demands, family support services are planned using PGIS	*Family resource mapping*: examines how to map and assess the spatial accessibility of family resources using PGIS applications *Analysis of housing and neighborhoods*: considers the use of PGIS in order to gain an understanding of the spatial features of neighborhood dynamics and family housing *Community-based family programs*: collectively, the community is involved in the creation of family support programs through the use of PGIS *Accessibility to services*: by utilizing PGIS, the evaluation and improvement of families' spatial access to essential services and assistance can be accomplished

(continued)

Table 10.1 (continued)

Sr. no.	PGIS and social sciences disciplines	Disciplinary principles	Research scope
10.2.16	PGIS in gender studies	*Gendered space analysis*: maps and analyzes the spatial aspects of gendered encounters using PGIS *Perspectives on feminist geography*: uses PGIS to analyze the geographical dimensions of power, identity, and social justice in gender studies *Community involvement*: promotes gender-neutral, active engagement of all genders in PGIS data collecting and interpretation procedures *Intersectionality*: uses PGIS to investigate how gender interacts with other social variables to shape spatial experiences	*Gender-based violence mapping*: examines how PGIS might be used to map and address spatial patterns of gender-based violence *Access to resources*: examines geographic differences in resource access according to gender using PGIS *Safe spaces mapping*: maps and creates safe spaces within communities using PGIS, involving a range of genders *Participatory urban planning*: utilizing PGIS, participatory urban planning involves the entire community in gender-responsive urban design
10.2.17	PGIS in geography	*Physical geography* *Climate analysis*: uses PGIS to analyze spatial patterns and fluctuations in climate *Landform mapping*: maps and analyzes spatial patterns in actual landforms using PGIS *Ecosystem dynamics*: maps and tracks spatial changes in ecosystems and biodiversity using PGIS *Natural hazard assessment*: maps and evaluates spatial risks and vulnerabilities to natural hazards using PGIS	*Physical geography* *Mapping geomorphic processes*: Examines how to map and analyze spatial patterns in geomorphic processes using PGIS *Spatial hydrology analysis*: analyzes spatial patterns in hydrological processes by exploring PGIS via the lens of spatial hydrology *Community-based environmental monitoring*: makes use of PGIS to facilitate cooperative community participation in the tracking and mapping of environmental changes *Land use planning*: sustainable land use planning and resource management based on spatial considerations are achieved through including communities in PGIS-based land use planning
		Economic geography *Mapping trade routes*: this technique makes use of PGIS to map and analyze trade routes and economic connections *Market accessibility*: maps and analyzes spatial patterns of market accessibility using PGIS *Supply chain optimization*: uses PGIS to optimize supply chain networks *Regional development planning*: uses PGIS to plan and carry out regional economic development initiatives	*Economic geography* *Economic disparities mapping*: uses PGIS to map and analyze geographical patterns of economic disparities within areas *Spatial market analysis*: Investigates PGIS applications for mapping and analyzing spatial dynamics in local and global markets *Supply chain resilience*: employing PGIS, communities may optimize supply chain resilience and efficiency depending on spatial constraints *Community-based business planning*: utilizes PGIS to facilitate collaborative community participation in local business planning

Social geography	*Social geography*
Social space analysis: uses PGIS to map and analyze the spatial dimensions of social interactions and systems *Identity and place*: uses PGIS to examine geographical dimensions of identity, belonging, and cultural expression *Community involvement*: encourages communities to actively participate in PGIS data gathering and interpretation *Spatial justice*: uses PGIS to investigate spatial inequities and advocate for social justice in geographical situations	*Social infrastructure mapping*: communities use PGIS to map and improve spatial access to social services and infrastructure *Community identity mapping*: investigates PGIS applications for mapping and analyzing the geographical elements of community identity *Participatory urban planning*: uses PGIS to engage the entire community in developing socially just urban areas *Marginalized community empowerment*: investigates the use of PGIS to identify and remediate spatial inequities that affect marginalized populations
Cultural geography	*Cultural geography*
Geographical ethnography: applies PGIS to examine geographical dimensions of cultural behaviours, rituals, and manifestations *Cultural landscape mapping*: applications of PGIS to map and analyze the spatial dimensions of cultural landscapes *Cultural heritage preservation*: employ PGIS to document and preserve spatial components of cultural heritage *Community involvement*: encourages communities to actively participate in PGIS data gathering and interpretation	*Cultural mapping*: investigate PGIS applications for mapping and analyzing spatial patterns in cultural phenomena *Cultural identification mapping*: investigates the use of PGIS to map and understand the spatial dimensions of culture identification *Cultural impact assessment*: engages communities in PGIS to examine the spatial impact of cultural practices and policies on cultural spaces *Community-based resources management*: PGIS is used to facilitate collaborative community participation in the preservation and management of cultural resources

(continued)

Table 10.1 (continued)

Sr. no.	PGIS and social sciences disciplines	Disciplinary principles	Research scope
		Political geography	*Political geography*
		Geopolitical analysis: uses PGIS to investigate the geographical dimensions of power dynamics, political conflicts, and geopolitical relationships *Political boundaries mapping:* PGIS is used for the mapping and analysis of the spatial dimensions of political territories and boundaries *Community involvement:* promotes communities' active involvement in PGIS-based procedures for gathering and interpreting political data *Spatial justice advocacy:* PGIS is used in spatial justice advocacy to investigate and resolve spatial injustices in political settings	*Political representation mapping:* examines how to map and analyze spatial trends in political representation using PGIS technologies *Spatial analysis of political conflicts:* examines PGIS for the purpose of studying the spatial aspects of political disputes and their settlements *Community-based political planning:* community-based political planning involves the cooperative community's participation in political decision-making through the use of PGIS *Advocacy for spatial equality:* communities in PGIS are involved in the advocacy of equal political representation and spatial justice
		Technical geography	*Technical geography*
		Data precision and accuracy: highlights the significance of exact and accurate spatial data for technical applications *Spatial data infrastructure:* PGIS is integrated with the spatial data infrastructure to enable effective data sharing and management *Geospatial technology integration:* PGIS is the primary tool used in geospatial technology integration, which is the process of mapping and spatial analysis *Applications for remote sensing:* combines PGIS with technology for remote sensing to provide a thorough geographic analysis	*Spatial data standardization:* standardized spatial data format development and implementation: examines PGIS applications for this purpose *Integrated technology solutions:* engages PGIS communities in the development of integrated solutions to technical problems in the administration and analysis of spatial data *Training in geospatial technologies:* makes use of PGIS to facilitate cooperative community education in geospatial technologies *Planning infrastructure:* examines PGIS as a tool for organizing and maximizing the development of spatial infrastructure

10.2.18	PGIS in gerontology	*Age-friendly space mapping*: this technique makes use of PGIS to map and analyze spatial dimensions that improve senior citizens' quality of life *Quality of life analysis*: uses PGIS to investigate geographical factors influencing older individuals' quality of life in their communities *Community involvement*: promotes senior citizens' active involvement in PGIS data collecting and interpretation procedures *Planning for accessibility*: uses PGIS to optimize spatial accessibility for senior citizens	*Age-friendly urban planning*: examines how to map and enhance the spatial elements of urban environments that are accessible to senior citizens using PGIS technologies *Social connectivity mapping*: the goal of social connectivity mapping is to encourage healthy aging by involving older persons in PGIS for the purpose of mapping and fostering social relationships within communities *Healthcare accessibility*: makes use of PGIS to evaluate and improve the elderly's geographical access to healthcare services *Community involvement for elderly services*: looks into PGIS for cooperative community involvement in the development and improvement of senior population services
10.2.19	PGIS in global studies	*Global connectedness mapping*: maps and analyzes spatial patterns in global connectedness using PGIS *Geopolitical analysis*: applications of PGIS to investigate the geographical dimensions of international power systems, geopolitical linkages, and wars *Cross-cultural collaboration*: promotes the active involvement of many international populations in PGIS data collection and interpretation procedures *Global justice advocacy*: uses PGIS to investigate and resolve spatial disparities in global settings	*Mapping transnational migration*: examines PGIS uses in mapping and examining spatial trends related to migration across borders *Cultural exchange mapping*: utilize PGIS to map and support cross-border cultural exchange activities *Global resource management*: makes use of PGIS to facilitate cooperative community engagement in globally sustainable resource management *Planning for humanitarian supplies*: considers PGIS to plan and maximize the distribution of supplies during international emergencies
10.2.20	PGIS in health studies	*Inclusivity*: engages a wide range of stakeholders to get thorough data *Empowerment*: empowers local communities with GIS tools *Transparency*: communicates methodology and findings clearly *Ethical considerations*: prioritizes privacy and secrecy *Capacity building*: increase local capacity for ongoing participation	*Community health mapping*: increases your grasp of local health dynamics *Spatial analysis*: investigates the spatial patterns of health issues *Community empowerment*: enables communities to solve health issues *Resource allocation*: facilitates the efficient allocation of health resources *Policy informatics*: gathers community feedback to inform evidence-based health policies

(continued)

Table 10.1 (continued)

Sr. no.	PGIS and social sciences disciplines	Disciplinary principles	Research scope
10.2.21	PGIS in history	*Oral history integration*: combines GIS with oral histories to develop richer narratives *Community engagement*: involves local communities in the mapping of historical places *Heritage conservation*: supports cultural heritage conservation with GIS *Archival mapping*: archival mapping involves digitizing and mapping historical materials to provide spatial context *Interdisciplinary collaboration*: collaborates with historians, archaeologists, and geographers to gain comprehensive perspectives	*Community heritage mapping*: getting communities involved in mapping local historical assets *Historical landscape reconstruction*: employs GIS to rebuild historical landscapes *Cultural heritage tourist planning*: plans heritage tourist routes based on participative GIS insights *Spatial analysis of events*: examines historical events in their spatial context *Archaeological site documentation*: Records and maps archaeological sites for preservation
10.2.22	PGIS in human services	*Client-centric mapping*: involves service receivers in mapping their requirements *Feedback integration*: incorporates client feedback for continuous improvement *Privacy and confidentiality*: when dealing with sensitive data, put privacy first *Equity focus*: addressing social equity challenges in service delivery *Accessibility*: makes sure PGIS tools are easy to use for a variety of service consumers	*Vulnerability mapping*: constructs a map of vulnerable groups to help them receive targeted interventions and assistance *Community asset mapping*: identifies and utilizes local assets for service delivery *Resource allocation optimization*: plans resource distribution based on spatial demand *Accessibility planning*: develops services to promote equal geographical accessibility *Needs assessment mapping*: uses PGIS to assess community needs in human services
10.2.23	PGIS in intelligence studies	*Collaborative analysis*: encourages teamwork among intelligence analysts that use PGIS *Continuous training*: provides continual training in PGIS technologies and procedures to intelligence professionals *Data security*: builds up strong security procedures for critical geospatial data *Interagency cooperation*: encourages information sharing and collaboration among agencies *Ethical considerations*: employs ethical standards when collecting and analyzing intelligence data	*Threat mapping*: utilizes PGIS to conduct geographical analyses of security threats *Surveillance planning*: makes use of PGIS to plan and optimize surveillance activities *Risk assessment*: evaluates and maps hazards to vital infrastructure *Crisis response planning*: creates PGIS-based strategies for timely crisis response *Geospatial pattern analysis*: identifies patterns and trends to gain strategic intelligence insights

10.2.24	PGIS in international relations	*Cross-border collaboration:* makes use of PGIS to foster collaboration across borders *Environmental diplomacy:* employs PGIS to promote environmental cooperation and diplomacy *Diplomatic transparency:* makes international ties more transparent by using PGIS data *Human rights integration:* incorporates human rights issues to PGIS analyses	*Environmental issues:* employing PGIS, investigate and address transboundary environmental concerns *Global health mapping:* visualizes and evaluates health issues with global ramifications *Conflict analysis:* uses PGIS to conduct spatial analyses of international conflicts *Diplomatic resource allocation:* with PGIS, you may optimize resource allocation for diplomatic operations *Migration studies:* investigates and maps worldwide migration patterns
10.2.25	PGIS in labour studies	*Equity considerations:* addresses spatial discrepancies in working circumstances *Worker participation:* engages employees in mapping workplace concerns and needs *Privacy protection:* protects worker privacy when processing geographical data *Collaborative bargaining:* encourages participation in PGIS-based labor-management negotiations	*Commute analysis:* investigates commuting patterns and their effects on labor *Workplace mapping:* examines the geographic dimensions of workplace settings *Union activity mapping:* determines and analyzes the spatial distribution of union activity *Job accessibility:* assesses the geographical accessibility of job opportunities
10.2.26	PGIS in legal studies	*Evidence support:* uses PGIS to provide spatially based legal evidence *Privacy protection:* when mapping legal data, ensures that you follow all privacy laws *Community input:* integrates community opinions into legal mapping projects *Access to justice:* uses PGIS to improve geographic access to legal services *Ethical considerations:* maintains ethical standards in the collecting and use of geospatial legal data	*Crime mapping:* uses PGIS to study and map criminal patterns *Land use regulation:* determines the geographical implications of legal land use rules *Access to legal resources:* maps and improves access to legal services *Environmental law compliance:* keeps track of spatial compliance with environmental standards *Community legal empowerment:* employs PGIS to empower communities in legal matters

(continued)

Table 10.1 (continued)

Sr. no.	PGIS and social sciences disciplines	Disciplinary principles	Research scope
10.2.27	PGIS in media studies	*Audience engagement mapping*: involves audiences in mapping their media preferences and consumption *Cultural representation*: ensures that varied cultures are fairly represented in media maps *Privacy protection*: when evaluating geographical data related to the media, keep individual privacy in mind *Collaborative storytelling*: enables the collaborative generation of geographic narratives in media initiatives. *Ethical content mapping*: uses ethical guidelines while mapping sensitive media content	*Media use patterns*: examine the spatial patterns of media use and preferences. *Cultural media impact*: investigates the regional influence of media on cultural perceptions *Audience input mapping*: uses PGIS to map and evaluate audience input and sentiment *Media accessibility mapping*: determines and maps the accessibility of media content across areas
10.2.28	PGIS in military science	*Operational security*: when managing classified geospatial military data, put security foremost *Ethical use of force*: observes ethical norms while using geographic information in military operations *Interagency collaboration*: PGIS can help military branches collaborate more effectively *Strategic planning assistance*: offers PGIS tools for strategic military planning *Continuous training*: ensures that military personnel receive continual training in PGIS technologies	*Terrain study*: in military operations, use PGIS to conduct extensive terrain studies *Logistics optimization*: optimizes military logistics using PGIS-based planning *Threat mapping*: visualizes and evaluates spatial trends in security threats *Humanitarian assistance planning*: uses PGIS to plan and implement military humanitarian assistance activities *Battlefield visualization*: improves situational awareness using PGIS-based battlefield visualization
10.2.29	PGIS in peace and conflict studies	*Inclusive conflict mapping*: involves multiple stakeholders in mapping conflict processes *Conflict-sensitive data handling*: manages sensitive data responsibly and with conflict awareness *Community involvement in peace building*: the mapping of peace building programs should involve community participation *Human rights integration*: human rights concerns need to be incorporated into the GIS study of conflicts *Transparency and trust-building*: encourages transparency in data sharing to foster trust among participants in peace initiatives	*Conflict hotspot mapping*: using PGIS, locate and map locations prone to conflict *Conflict resolution planning*: uses PGIS to plan conflict resolution methods and interventions *Peace building resource allocation*: using PGIS, optimize resource allocation for peace building projects *Refugee and displacement analysis*: identifies and maps patterns of refugee flow and displacement *Post-conflict reconstruction*: it is important to map and organize the reconstruction efforts that will take place in post-conflict zones in order to ensure peace for the long term

10.2.30	PGIS in political science	*Citizen engagement*: involves citizens in mapping their political preferences and issues *Data transparency*: keeps political data collection and mapping as transparent as possible *Collaborative policy mapping*: enables collaborative mapping to support political decision-making *Privacy protection*: maintains individual privacy in politically sensitive PGIS projects *Representation equity*: aims for fair and equal representation in political PGIS analysis	*Electoral districting analysis*: utilizes PGIS to assess and optimize electoral district boundaries *Political participation mapping*: studies the spatial dimensions of public engagement and political participation *Political opinion mapping*: visualizes and analyzes spatial patterns in political attitudes *Resource allocation for governance*: using PGIS, optimize resource distribution for effective governance *Corruption hotspot analysis*: using PGIS, detect and address spatial patterns of political corruption
10.2.31	PGIS in psychology	*Informed consent*: prioritizes informed consent when mapping psychological data *Community collaboration*: engages communities in mapping mental health resources and issues *Privacy protection*: in psychological PGIS initiatives, individuals' privacy and confidentiality are protected *Cultural sensitivity*: cultural sensitivity is important to consider while mapping psychological processes *Ethical data handling*: when collecting and analyzing psychological geographic data, follow ethical norms	*Mental health resource mapping*: uses PGIS to map and analyze the distribution of mental healthcare resources *Environmental impact on mental health*: looks into the geographic effects of environmental elements on mental health *Social network analysis*: uses PGIS to examine spatial trends in social connections and support networks *Community well-being assessment*: engages communities in mapping elements that influence psychological well-being *Access to psychological services*: evaluates the geographical accessibility to psychological services
10.2.32	PGIS in public administration	*Citizen engagement*: involves citizens in mapping their public service needs and preferences *Equal service allocation*: uses PGIS to ensure the fair and equal allocation of public services *Data transparency*: maintains transparency in the gathering and use of geographic public administration data *Collaborative policy planning*: enables stakeholders to collaborate and make informed public policy decisions	*Service accessibility analysis*: uses PGIS to assess and improve the geographic accessibility of public services *Community development mapping*: gets communities involved in mapping out focused development activities *Disaster response planning*: employs PGIS to plan disaster responses and recoveries more efficiently *Urban planning and infrastructure*: incorporates PGIS to optimize urban planning and infrastructure development *E-governance implementation*: integrates PGIS to ensure effective e-governance projects and citizen participation

(continued)

Table 10.1 (continued)

Sr. no.	PGIS and social sciences disciplines	Disciplinary principles	Research scope
10.2.33	PGIS in religious studies	*Sacred site mapping:* identifies and maintains religiously significant sites *Interfaith collaboration:* encourages collaboration between different religious congregations on PGIS projects *Cultural sensitivity:* involves varied religious perspectives in mapping efforts *Privacy considerations:* keeps sensitive religious data with the utmost privacy and confidentiality *Ethical data representation:* ensures that religious data is ethically represented in geographical analytics	*Pilgrimage route analysis:* uses PGIS to evaluate and optimize pilgrimage routes *Sacred landscape exploration:* studies the spatial distribution and interconnections of sacred landscapes *Interfaith dialogue support:* incorporates PGIS in programs that promote understanding and dialogue across different religious communities *Religious demographics mapping:* visualizes and studies the spread of religious communities *Cultural heritage preservation:* uses PGIS to preserve religious and cultural heritage places
10.2.34	PGIS in rural studies	*Local knowledge integration:* uses indigenous and local knowledge in PGIS analysis *Community participation:* engages rural communities in mapping local needs and resources *Equitable resource distribution:* uses PGIS to ensure a fair and equitable distribution of resources *Sustainable development:* uses PGIS to plan and promote sustainable rural development	*Agricultural land usage mapping:* examines and improves spatial patterns of agricultural land usage *Community-based disaster preparedness:* identifies and plans for disaster response in rural areas *Access to basic services:* evaluates the geographical accessibility of critical services in rural areas *Natural resource management:* employs PGIS to ensure that rural natural resources are managed sustainably *Livelihood diversification analysis:* analyzes spatial trends in diversifying rural livelihoods to increase economic resilience
10.2.35	PGIS in science and technology studies	*Technology impact mapping:* technology impact mapping examines the spatial impact of technology on society *Accessibility considerations:* PGIS can help ensure fair access to technical resources *Public participation in innovation:* engages the public in mapping and guiding technical developments *Ethical data handling:* when gathering and using geospatial technology data, follow ethical guidelines *Interdisciplinary collaboration:* work with people from many fields to gain a broad grasp of how technology affects society	*Environmental impact assessment:* uses PGIS to determine the environmental impact of technological advancements *Digital divide analysis:* investigates spatial patterns in the digital divide and access to technology *Technology adoption patterns:* analyzes the spatial variations in the adoption of developing technologies *Science communication mapping:* visualizes and evaluates the efficacy of science communication projects

10.2.36	PGIS in social justice studies	*Inclusive mapping:* gets underprivileged communities involved in mapping social inequalities *Advocacy and empowerment:* uses PGIS to advocate and empower marginalized populations *Ethical data handling:* follows ethical principles when gathering and using geographic data relevant to social justice *Collaboration with activists:* works with social justice activists to create educated and effective PGIS initiatives	*Environmental justice mapping:* investigates the geographical dimensions of environmental justice and its impact on communities *Spatial inequality analysis:* uses PGIS to map and examine spatial patterns of social inequality *Access to social services:* determines the geographical differences in access to critical social services *Community policing analysis:* determines the spatial impact of policing policies on underprivileged groups *Intersectionality mapping:* mapping overlapping forms of injustice allows you to investigate the intersectionality of social concerns
10.2.37	PGIS in social policy	*Stakeholder engagement:* collaborates with a variety of stakeholders to map social policy needs *Equitable resource allocation:* uses PGIS to ensure a fair and equitable distribution of resources *Data transparency:* ensures that social policy data is handled and communicated transparently *Collaborative decision-making:* enables collaborative mapping to inform social policy decisions	*Welfare program impact analysis:* uses PGIS to determine the spatial impact of social welfare initiatives *Community well-being assessment:* involves communities in mapping issues that influence overall well-being to inform successful policy planning *Education and job access mapping:* determines and maps access to educational and job options *Healthcare resource optimization:* improves healthcare resource distribution to achieve better social policy outcomes *Housing and urban development planning:* employs PGIS to plan and evaluate housing and urban development policy
10.2.38	PGIS in social work	*Cultural competence:* during developing PGIS projects for social work, keep cultural nuances in mind *Client empowerment:* enables clients to map their needs and resources *Ethical data handling:* though gathering and using geographic data, follow ethical guidelines *Collaborative case management:* allows stakeholders to work together to manage cases more effectively	*Vulnerability analysis:* analyzes spatial patterns of vulnerability to help guide social work interventions *Community resource mapping:* uses PGIS to map and evaluate accessible community resources *Access to social services:* determines geographic differences in access to social services *Disaster response planning:* uses PGIS to map vulnerabilities and develop effective disaster response strategies

(continued)

Table 10.1 (continued)

Sr. no.	PGIS and social sciences disciplines	Disciplinary principles	Research scope
10.2.39	PGIS in socio-ecological studies	*Ethical environmental data handling:* when collecting and analyzing socio-ecological data, follow ethical guidelines *Community participation:* involves communities in mapping socio-ecological systems *Ecosystem diversity preservation:* uses PGIS to map and conserve biodiversity in ecosystems *Interdisciplinary collaboration:* works with researchers from several fields to gain broad socio-ecological insights *Sustainable resource management:* promotes PGIS-based technologies for the sustainable management of ecological resources	*Climate change impact analysis:* examines the spatial implications of climate change for socio-ecological systems *Biodiversity mapping:* employs PGIS to map and monitor biodiversity patterns *Land use change analysis:* investigates and analyzes spatial patterns of land use change in socio-ecological systems *Ecosystem service assessment:* uses PGIS to evaluate and map the delivery of ecosystem services *Community-based conservation planning:* collaborates with local communities to lay out long-term conservation initiatives
10.2.40	PGIS in sociology	*Community participation:* gets communities involved in mapping sociological phenomena *Privacy protection:* protects individual privacy and confidentiality in sociological PGIS projects *Cultural sensitivity:* takes cultural nuances into account when working on sociology-related PGIS projects *Collaborative research:* facilitates collaboration among stakeholders to gain complete sociological insights	*Urban sociology and spatial patterns:* studies the spatial dynamics of urban social structures and interactions *Social inequality mapping:* applies PGIS to map and examine spatial patterns of social inequality *Community network analysis:* studies and maps the social networks inside communities *Migration and cultural integration mapping:* visualizes and investigates the spatial dimensions of migration and cultural integration *Community resilience mapping:* utilizes PGIS to measure and map community resilience characteristics in sociological contexts
10.2.41	PGIS in urban studies	*Community engagement:* engages urban inhabitants in mapping local needs and resources *Equal resource allocation:* employs PGIS to ensure a fair and equal distribution of urban resources *Sustainable development:* using PGIS to plan and promote sustainable urban development *Ethical data handling:* during collecting and analyzing geospatial urban data, follow ethical guidelines	*Smart city development:* uses PGIS to design and execute smart city initiatives *Transportation planning:* employs PGIS to optimize urban transportation networks *Spatial inequality analysis:* identifies and addresses spatial patterns of inequality in metropolitan areas *Disaster resilience planning:* makes use of PGIS to map risks and design disaster-resistant urban infrastructure

demonstrate the organization's dedication to honoring multiple perspectives and guaranteeing equal representation. Furthermore, participatory GIS research has a substantial impact on the social sciences by tackling topics such as social inequity, community networks, resource distribution, and disaster response planning. It acts as a catalyst for educated decision-making, promoting social justice, sustainable development, and community empowerment. Integrating participatory GIS into the social sciences improves research quality and relevance, fostering a more inclusive and equitable approach to tackling societal concerns.

Further Reading

Brown, G., & Kyttä, M. (2014). Key issues and research priorities for public participation GIS (PPGIS): A synthesis based on empirical research. *Applied Geography, 46*, 122–136. https://doi.org/10.1016/j.apgeog.2013.11.004

Ndzabandzaba, C. (2019). Participatory geographic information system (PGIS): A discourse toward a solution to traditional GIS challenges. In S. Brunn & R. Kehrein (Eds.), *Handbook of the changing world language map*. Springer. https://doi.org/10.1007/978-3-319-73400-2_122-1

Oyana, T. J. (2017). The use of GIS/GPS and spatial analyses in community-based participatory research. In S. S. Coughlin, S. A. Smith, & M. E. Fernandez (Eds.), *Handbook of community-based participatory research*. Oxford Academic. https://doi.org/10.1093/acprof:oso/9780190652234.003.0004

Parker, R. N., & Asencio, E. K. (2009). *GIS and spatial analysis for the social sciences: Coding, mapping, and modeling*. Routledge. ISBN: 9781135857585.

Ramasubramanian, L. (2008). *Geographic information science and public participation* (Advances in geographic information science). Springer. https://doi.org/10.1007/978-3-540-75401-5. ISBN: 978-3-540-75,401-5.

Steinberg, S. J., & Steinberg, S. L. (2005). *Geographic information systems for the social sciences: Investigating space and place*. Sage Publications. https://doi.org/10.4135/9781452239811. ISBN: 9780761928737.

Author Contributions

Book conceptualization: K.D.M.; Project developed, correspondence and management: K.D.M.; Lead Author and supervision: K.D.M.; Co-Author: S.R.; Field investigation and data calculation: S.R. and K.D.M.; Field narratives thoughts: S.R. and K.D.M.; Methodologies and software: K.D.M. and S.R.; Tables, figures, photo plates, and mapping: S.R. and K.D.M.; Resources, Research and Writing: K.D.M. and S.R.; Reviewing Roles: K.D.M., S.R., and other anonymous reviewers; Editing: K.D.M. and S.R.; Original draft design and preparation: K.D.M. and S.R.; Proofing and others: K.D.M. and S.R.

K.D.M.: Kousik Das Malakar; S.R.: Supriya Roy

K. D. Malakar, S. Roy, *Mapping Geospatial Citizenship*, SpringerBriefs in GIS,
https://doi.org/10.1007/978-3-031-63107-8

Glossary

Asset mapping   It is the process of identifying and mapping community assets such as natural resources, infrastructure, and social networks.

Capacity building   Emphasizing the need to provide communities and individuals with the required skills and knowledge to participate actively in PGIS programs.

Citizen science   Involvement of the general public in scientific research, including geospatial data collection and analysis.

Climate resilience   Investigative techniques and measures to improve communities' ability to adapt to and recover from the effects of climate change.

Cognitive mapping   It is an emotional depiction of space and place that is frequently used to better understand how people perceive and navigate their surroundings.

Community cartography   The collaborative process of creating maps with active participation from community members, with an emphasis on local knowledge and opinions, is referred to as community cartography.

Community engagement   Community engagement is the participation of members of a community in decision-making processes and activities that influence their lives and surroundings.

Community mapping   The process of making maps with active participation from members of the community, frequently to represent local traditional knowledge, technical knowledge, resources, or challenges.

Community-based conservation   Emphasizing the critical function of PGIS in empowering local communities to take an active role in environmental management and conservation.

Critical cartography   It is the construction of maps that question traditional perspectives, power systems, and spatial representations.

Critical GIS   A way of questioning the assumptions and power structures underlying traditional GIS techniques, with a focus on social justice and empowerment.

Cultural mapping   The practice of documenting and portraying a community's cultural legacy through maps and spatial data.

K. D. Malakar, S. Roy, *Mapping Geospatial Citizenship*, SpringerBriefs in GIS,
https://doi.org/10.1007/978-3-031-63107-8

Data integration   It is the study of the fusion of multiple kinds of data kinds, including both community wisdom and scientific information, in order to assist well-informed decision-making processes.

Data literacy   Citizens involved in geographic issues should be able to interpret and critically examine geospatial data, ensuring its precision and relevance.

Digital divide   The gap between individuals who have access to contemporary information and communication technology and those who do not, which is typically tied to socioeconomic considerations, is known as the digital divide.

Ecological knowledge   It is the information and understanding about the ecosystems, environment, and natural processes that is commonly held by local community populations.

Eco-mapping   It is a type of mapping that focuses on depicting ecological characteristics, processes, and relationships in a specific area.

Ecosystem services   Emphasis on the benefits that ecosystems provide to people and how PGIS may help with the evaluation and long-term management of these services.

Environmental justice   This topic will discuss the concept of equity in the distribution of environmental rights and responsibilities, as well as how PGIS can be utilized to promote justice.

Fishing grounds   Identifying specific geographic locations where fishing operations occur, which are frequently necessary for subsistence.

Geospatial citizenship   The ethical and knowledgeable use of geospatial technologies and data for comprehending the world, making decisions, and engaging in civic life is known as geospatial citizenship.

Geospatial education   Emphasizing the need of deliver high-quality GIS training and education to equip people with the skills they need to engage in a geospatially enabled world.

Geospatial equity   Emphasizing the importance of providing all persons and communities with fair and equal access to geospatial technologies and information.

GIS (geographic information systems)   GIS is a critical tool in geospatial citizenship because it allows for the collection, analysis, and visualization of geographic data.

GIS facilitator   A person or group that promotes and directs the participatory GIS process, encouraging communication and the sharing of knowledge.

GIS for social justice   The application of geographic information systems (GIS) technologies to solve social disparities, advocate for justice, and empower underprivileged populations.

GIS toolkit   A collection of information, tools, and procedures used in GIS initiatives, sometimes tailored to the needs and circumstances of a certain community.

Inclusive GIS   The technique of making GIS tools, data, and applications available to a wide range of user groups, independent of experience or skill.

Indigenous technical knowledge   Emphasizing the distinct traditional knowledge and practices owned by a place's or space's local or traditional communities.

Local knowledge   Talking about how important it is to include knowledge from the indigenous people or the community when mapping.

Map design principles   Guidelines for producing effective and communicative maps, taking into account elements such as scale, symbols, and colour selection.

Map literacy   It is the process by which communities acquire the information and abilities necessary to comprehend, analyze, and utilize maps as instruments for dialogue and decision-making.

Mapathon   An event in which a group of people collaborate on mapping projects, typically utilizing online mapping platforms.

Mobile GIS   This technology improves the accuracy and accessibility of spatial data for a range of applications by enabling real-time geographic data collection, analysis, and management using portable devices such as smartphones and tablets.

Participatory action research   A method of addressing social issues that involves people in the research process by combining research and action.

Participatory decision-making   It includes actively engaging individuals and stakeholders in collaborative choices, encouraging inclusivity, transparency, and shared responsibility to address difficult issues and achieve collective goals.

Participatory geographic information system (PGIS)   A cooperative approach that incorporates neighborhood groups in the creation, evaluation, and interpretation of geographic data is known as participatory GIS.

Participatory mapping   It is the process of collaboratively creating maps that reflect the priorities and spatial knowledge of community members.

Participatory planning   It entails involving local communities in planning, development, and resource management decision-making processes.

Public participation   Involving community members in decision-making processes to ensure their perspectives are heard and considered.

Qualitative GIS   Integrating qualitative data, such as narratives and personal experiences, with spatial information to create a more complete knowledge of a location.

Remote sensing   The technology and methods used to collect data from a distance, typically from satellites or aerial platforms, for monitoring and analysis.

Resilience strategies   Emphasis on methods and interventions that can improve mangrove ecosystem resilience in the face of climate change impacts.

Resource management   Examining how enhanced management and preservation of coastal fishing resources are made possible by participatory mapping.

Risk assessment   It is centered on assessing the community's possible exposure to climate-related risks and vulnerabilities, frequently with the aid of geospatial data and tools.

Social mapping   The use of maps to portray social aspects, interactions, and dynamics, which is commonly done cooperatively with community members.

Socio-ecological mapping   Socio-ecological mapping visualizes and analyzes complex relationships between human societies and ecosystems, enabling sustainable management and policy decisions by showing environmental and societal interactions.

Socio-ecological system   An approach that addresses the linkages and interdependence of a particular environment's social and ecological components.

Socio-spatial ecologies   This word emphasizes the connectivity of social and spatial aspects within a certain societal place/space and how they influence one another.

Space-based knowledge   Emphasizing the importance of local knowledge and views in recognizing and addressing the unique issues and possibilities that exist within a particular sociocultural space.

Spatial analysis   It is the process of looking for patterns and relationships in spatial data in order to obtain insights and make informed decisions.

Spatial empowerment   Discussing how community cartography and geographic information systems (GIS) enable individuals and communities to become active participants in designing their environments and addressing local challenges.

Spatial equity   It refers to the equitable allocation of resources, opportunities, and benefits across geographic areas or populations.

Spatial justice   The concept of guaranteeing fair access to and benefits from geographic information and technology, hence addressing geospatial citizenship imbalances.

Spatial knowledge   Information regarding a community's spatial characteristics, such as local landmarks, resources, and cultural or historical locations.

Spatial literacy   Understanding, interpreting, and communicating information about location, space, and spatial relationships is referred to as spatial literacy.

Spatial thinking   The ability to recognize and interpret spatial interactions is required for effective geospatial citizenship.

Spatially explicit modeling   It is the use of spatial data to build models that depict and evaluate the dynamics of socio-ecological systems.

Story mapping   Using maps to tell a story or describe a series of events, frequently with multimedia features.

Storytelling maps   Maps that include narratives and stories to provide a more in-depth understanding of a community's history, culture, or difficulties.

Volunteered geographic information (VGI)   Geographic data that is voluntarily supplied by individuals, frequently through crowdsourcing operations.

Web GIS   Web-based GIS accesses, analyzes, and shares geospatial data and maps using internet browsers. It improves web access, interactivity, collaborative mapping, and decision-making.

Index